T0258509

THE DIGITAL REVOLUTION

"Offers timely insight into a timeless preoccupation with the digital age."

Benjamin Peters
Hazel Rogers Associate Professor of Media
Studies and affiliated faculty Cyber Studies,
University of Tulsa, and author of *How Not to Network a Nation:*
The Uneasy History of the Soviet Internet

"Gabriele Balbi delves into a notion whose history, actors and developments shape our digital imaginaries and practices, as well as our relationship with technology, media and innovation. A must-read for anyone interested in the digital world."

Valérie Schafer
Center for Contemporary and Digital History,
University of Luxembourg

"This short book is both topical and timely."

Jane Winters
Professor of Digital Humanities,
School of Advanced Study, University of London

THE DIGITAL REVOLUTION

A Short History of an Ideology

Gabriele Balbi

Translated by
Bonnie McClellan-Broussard

Translation supported by Fondazione Hilda e Felice Vitali, Lugano

OXFORD
UNIVERSITY PRESS

OXFORD
UNIVERSITY PRESS

Great Clarendon Street, Oxford, OX2 6DP,
United Kingdom

Oxford University Press is a department of the University of Oxford.
It furthers the University's objective of excellence in research, scholarship,
and education by publishing worldwide. Oxford is a registered trade mark of
Oxford University Press in the UK and in certain other countries

A previous version of this book was published in Italian in 2022: *L'ultima ideologia.
Breve storia della rivoluzione digitale* (Laterza).

Published in the United States of America by Oxford University Press
198 Madison Avenue, New York, NY 10016, United States of America

British Library Cataloguing in Publication Data

Data available

Library of Congress Control Number: 2023937659

ISBN 9780198875970

DOI: 10.1093/oso/9780198875970.001.0001

Printed and bound in the UK by
Clays Ltd, Elcograf S.p.A.

Contents

Foreword

Vincent Mosco

In this short but insightful book, Dr. Gabriele Balbi tells the story of an idea that has captured the imagination as much as or more than any other since the middle of the last century. It is a tale that asserts, often forcefully, that we are in the midst of a computer revolution that constitutes a radical rupture from the past and which irrevocably changes the lives of those living through it and future generation as well. Specifically, the transition from an analog to a digital world is not just a shift in technology, it portends a fundamental change in time, space, and social relations.

Having spent the past 50 years engaging with the story Dr. Balbi recounts, I am familiar with the range of ideas and emotions it conjures. In *The Digital Sublime* (Mosco, 2004), building on the work of writers like Leo Marx (1964), David Nye (1996), and Carolyn Marvin (1990), I made it the object of a book that describes the idea of a digital revolution as more than just a change in technology and even more than a new set of political economic arrangements. Just as importantly, it marks the latest iteration in the grand myths that provide fodder for our emotional comfort and ideological justification for a changing world. In a word, it offers *transcendence* to humanity as it seeks to cope with the consciousness of its own mortality just as, in the past, the natural world and religion provided pathways that might take us beyond the limits of our biology. This latest vision of the technological sublime, now embodied in myths about artificial intelligence and the metaverse anticipates, for its major storytellers and believers, nothing short of the end of history, the end of geography, and the end of politics.

Dr. Balbi's analysis casts a wide net that includes supporters and critics of the digital revolution thesis to suggest that whether optimist or pessimist, believer or doubter, most experts have viewed the coming of the computer as a monumental event, a watershed in human history. On the surface, it is understandably hard to contend with the use of the term revolution to describe the arrival of a digital world. From measures of screen time to the power of companies like Google, Amazon, Microsoft, Apple, and Facebook that developed the applications that most of the globe depends on and which make them the wealthiest companies in the world, one can justifiably argue for the transcendent importance of the digital world. Nevertheless, when one considers other measures, such as questionable productivity gains and the unkept promises of a world where leisure replaces work, the word *banality* may seem more appropriate than transcendence. When thought about in connection with the capacity to deepen and extend surveillance, rather than freeing us from the "iron cage" of a bureaucratic society that the sociologist Max Weber (2002) foresaw over a century ago, the digital world tightens the bars and strengthens the locks.

One of the benefits of Dr. Balbi's book is that it encourages reflection on the digital revolution story in the context of the existential challenges the world faces today. Over the last three years, I have been working on a project that addresses the current resurgence of interest in the concept of utopia—a term given life in Thomas More's 1516 classic book (More, 2003). Leading today's renewed attention is a raft of new writing (from criticism to science fiction), art, music, and film on Afrofuturism, radical feminism, eco-socialism, and queer utopias. With all the talk about the metaverse and AI, there is interest in computer utopias, particularly of a libertarian bent, but these are secondary when compared to creative work on race, gender, social class, and the environment, especially when it comes to the most ambitious reimagining of what a better world might look like. Decolonizing utopia, rethinking gender and sexuality, revisioning a world where nature is pre-eminent, these are the primary themes in the worldwide renewal of interest in More's idea.

In my view, a key to understanding the resurgence of work on utopia, from the imaginations of people like Nora Keita Jemisin (2020, 2022), Kim Stanley Robinson (2020), Octavia Butler (2009), José Esteban Muñoz (2019), and John Bellamy Foster (2020), is the recognition that the world faces existential challenges that are beyond mere reforms. These include climate catastrophe, global pandemics, the risk of nuclear war, and the growing gap between rich and poor. Computers are implicated in all, both by deepening their risks and in improving the chances of human survival. However, none of these threats to the future of life on earth are simply the result of a digital revolution. They are the consequences of political and economic choices humans have made since the dawn of capitalism, especially among the most powerful people on the planet. The current fascination with utopia is largely an effort to apply the powers of human intelligence and imagination to rethink what it means to be human and to live in harmony with nature and our fellow citizens on this planet.

The revolutions we require are not first and foremost technological in nature. As Dr. Balbi reminds us, the sooner we dispose of linking revolution to whatever technology comes along, the more likely we are to address the revolutions in thought and in social relations that are required to save the planet from irreparable damage to the climate, from the potential of a pandemic far worse than COVID-19, from the likelihood of nuclear winter, and from the consequences of extreme differences in wealth and access to resources. Choices need to be made about how best to apply digital technologies but the solutions, as always, are to be found in how we decide to think about these problems and in how we choose to relate to one another.

Considering the stakes for the survival of our species and our planet, one might think that it would be reasonable to expect that setting aside the narrative of a grand digital revolution would be easy. However, it is one of the great strengths of Dr. Balbi's book that he convincingly documents the difficulties in doing so. The technological revolution he identifies is tantamount to a religion with its own mantras, faith system, and proud believers.

He identifies five primary mantras that provide sustenance to those craving digital transcendence. These include the widely used descriptive words: disruptive, total, irresistible, future, and permanent. The computer revolution disrupts a world in need of a shakeup. Its impact is total and resistance is futile. It is the wave, if not the tsunami, of the future and its impact is permanent.

In essence, as the book documents, the digital revolution is difficult to challenge because it is more than an intellectual stance and more than a theory whose veracity needs to be tested against the reality of everyday life today. It is, in essence, a religion with its own shrines, true believers, and heretics. As such it is difficult, if not impossible to displace. But all myths meet their adversaries, whether it is social movements that recognize the need to right wrongs or just a critical mass of individuals committed to undermining dominant myths and proposing more human-centered alternatives. Often, technological myths wither from what Weber also called the routinization of charisma or the banality that inevitably follows in the wake of a burst of religious enthusiasm. Such is likely to be the fate of the digital revolution, as discourse and as religion.

It is up to us, those struggling to find genuine solutions to address challenges to life on Earth to determine whether the story of a computer revolution, when it is relegated to the dustbin of history, will be replaced by something that might actually help save the world. Dr. Balbi's book is a good place to start.

References

Butler, Octavia. (2009). *Kindred*. Beacon.

Foster, John Bellamy. (2020). *The Return of Nature: Socialism and Ecology*. Monthly Review Press.

Jemisin, N.K. (2020). *The City We Became*. Hachette.

Jemisin, N.K. (2022). *The World We Made*. Orbit.

Marvin, C. (1990). *When Old Technologies Were New: Thinking about Electric Communication in the Late Nineteenth Century*. Oxford University Press.

Marx, L. (1964). *The Machine in the Garden*. Oxford University Press.

More, T. 2003. *Utopia*. Penguin.

Mosco, V. (2004). *The Digital Sublime*. The MIT Press.

Muñoz, J. E. (2019). *Cruising Utopia, 10th Anniversary Edition: The Then and There of Queer Futurity*. New York University Press.

Nye, D. (1996). *The American Technological Sublime*. The MIT Press.

Robinson, K.S. (2020). *The Ministry for the Future*. Orbit.

Weber, M. (2002.) *The Protestant Ethic and the "Spirit" of Capitalism and Other Writings* (P. Baehr and G. C. Wells, Trans.). Penguin.

Acknowledgements

This book is now starting its third life.

Its first life began with its conception in 2018, continued through early 2020, and culminated with a sabbatical granted by my university, USI Università della Svizzera italiana, in the Spring semester of 2020. As we all know, the first half of 2020 was also one of the most difficult moments in contemporary history for people around the globe, and I was only able to spend a few weeks of the time I had planned at Concordia University's Department of Communication Science in Montreal. Nevertheless, thanks to the support of Charles Acland, even in that brief period I was able to talk with several people in the Media History Research Centre about my book project. Afterwards, I also discussed my ideas with friends and colleagues over Skype and via an ever-increasing number of other platforms that were becoming popular at that time. In the first half of 2020, I remember illuminating chats with Bill Bruxton, Andreas Fickers, Richard R. John, Vincent Mosco, Katharina Niemeyer, Valérie Schafer, Dominque Trudel, Dwayne Winseck, and many others. Concordia University and Montreal itself, with its public libraries and stunning databases, were both crucial in the collection of sources.

In its second life, this book spoke Italian. During the long lockdowns, I had time to research the sources in-depth and to write the book slowly, in the Italian countryside (a very small village called Pollastra, in Piedmont) and in Lugano. While writing the book I also enjoyed fruitful conversations with Italian-speaking colleagues, most of whom were working at or connected to the Institute of Media and Journalism at my university. Some of them did me the service of reading over the manuscript. I am in debt to Deborah Barcella, Paolo Bory, Philip Di Salvo, Ely Lüethi, Paolo Magaudda, Simone Natale, Gianluigi Negro, Peppino Ortoleva, and Maria Rikitianskaia for their suggestions and revisions. Their help was crucial in refining my ideas, adding new historical examples, and changing my mind about some issues.

The book was published in February 2022 by the Italian publisher Laterza, thanks to the help of the editor Lia di Trapani. The volume immediately garnered some attention from the press and my academic colleagues. It has been presented and discussed at the universities of Arezzo, Siena, Cagliari, Udine, Lugano, Padua, Naples, Neuchâtel, Bari, Bologna, Turin, and several more are scheduled. For all of these kind invitations, I would like to thank Luca Barra, Gianenrico Bernasconi, Tiziano Bonini, Massimo Cerulo, Fausto Colombo, Marco Cucco, Juan Carlos De Martin, Sabino di Chio, Stefano Bory, Simone Dotto, Anna Maria Lorusso, Pietro Montorfani, Gianfranco Pecchinenda, Massimiliano Panarari, Francesca Pasquali, Stefano Pisu, Massimo Rospocher, Vito Saracino, Massimo Scaglioni, and Cosimo Marco Scarcelli, among many others.

The third life of *The Digital Revolution*, which is now in print, began in the Summer of 2022, when I decided to submit a book proposal to Oxford University Press. The editor Dan Taber immediately liked the idea, as did four anonymous reviewers. The latter also provided interesting insights, new scientific literature, and new case studies which helped me reshape the book for an international audience. Following my revision, Bonnie McClellan-Broussard started to work on translating the text, and she turned out to be much more than a simple translator. The book is very readable, at least to the untrained eye of a non-native speaker like myself, mainly because of her. The translation was sponsored by the Lugano-based Fondazione Hilda e Felice Vitali and the book would not have been possible without the foundation's generous support. The manuscript submission was then managed by Giulia Lipparini and finally Rajeswari Azayecoche who guided me through the production phase.

The last person I would like to thank is Francesco Balbi, who was born on December 2, 2020. Basically, he grew up together with the book and so it reminds me of the smell of milk, but also of his smiles and his first words, and even his first days of school. Franci, this book is dedicated to you.

Introduction
Understanding the Digital Revolution as an Ideology

We're often told how lucky we are and what a privilege it is to live in the era of the digital revolution. We all firmly believe that we're immersed in a historical moment of radical change for humanity, the effects of which are (or will be) as earth-shattering as those associated with the invention of fire, the French Revolution, or the Industrial Revolution. It is a revolution that can no longer be resisted because putting up resistance puts us at risk of being excluded from the sea of current and especially future opportunities. It is a revolution that we must believe in because—almost as if it were a religion—it will show us a path to follow. However, it is also a revolution that must be thoroughly understood because it is the foundation of some of contemporary society's most common dreams, expectations, and mythologies.

Without a doubt, digitization has changed and is changing the habits of billions of people around the globe. Creating and sharing photos or videos, sending and receiving messages, collecting and consuming information, buying and reselling objects on the Web—all of these have become daily and sometimes obsessive activities. It's a given that we own phones (on average, one for every person on Earth), which we keep with us for most of the day and which allow us to create still or moving images, to enjoy audiovisual content, to surf the Internet, to wake up in the morning, to play, and even to make phone calls. Those who don't have smartphones are seen as rare beasts: suspect. Discussing the alleged positive or negative "effects" of social media, 5G, and artificial intelligence (AI)—often expressing eccentric opinions—is a fascinating activity pursued by people at bus stops, in front of Apple stores, at dinners with friends at home, and among Facebook groups that predict digital apocalypses or liberations.

However, supporting or criticizing the theory that humanity is dealing with, or has dealt with, an evolutionary leap thanks to the digital "revolution" isn't the topic of this book. Nor do I intend to argue that this revolution is good or bad for humanity. That's not the point. Instead, this book analyzes the ways in which the great tale of the digital revolution has been *told*, the rhetoric, the narratives, and the overt or implied debates that have accompanied it and continue to accompany it today. In short, this volume aims to tell the *story of an idea*, probably the most powerful idea of recent decades: that digitization constitutes a revolution, a break with the past, a radical change for human beings who find themselves living through it.

The book aims to investigate the origins of this idea, how it evolved, which other past revolutions consciously or unconsciously inspired it, which great stories it has conveyed over time, which of its key elements have changed, and which ones have persisted and have been repeated in different historical periods. All of these discussions, large and small, have settled and become condensed into a series of media, advertising, corporate, political, and technical sources. I make extensive use of these sources in the following pages, sources that have resurfaced thanks to searches on digital databases such as Factiva, ProQuest, LexisNexis, Cartoon Stock, or have been collected over years of studies on the subject. I will also try to make these sources heard above the fray because I believe that the spirit of the digital revolution emerges more clearly from the actual words of politicians and corporate CEOs, from advertisements or cartoons, from media representations of the revolution, and from the thousands of paraphrases or interpretations that can be made of it. To tell the story of an idea, I believe that it is necessary to follow the traces that, voluntarily or involuntarily, it has left behind. The digital revolution has spilled rivers of ink and squandered words as well as expectations—political, economic, or social.

All these stories have an ideological quality. Specifically, an ideology—with many of the strong and weak associations that this term has assumed—that has proved capable of reaching across global societies between the second half of the twentieth century and the first decades of the twenty-first. A truly global ideology made up of discussions, which are surprisingly similar even in distant countries and cultures, of stereotypes repeated and renewed over time, of a linear vision of history that leads from an analog world to an inevitably digital one.

But why should the digital revolution be an ideology? The term ideology has essentially two meanings that have filtered into the common consciousness, the first being a specific vision of the world and the second a false or distorted way of thinking. Both of these dimensions are applicable to the digital revolution. Starting with the concept of ideology as a vision of the world, the digital revolution is an idea that has dominated recent decades. Having entered the collective awareness, it has made digitization the supreme principle, relegating everything else—the non-digital—to the background. It has a systematic character, that provides an exhaustive and coherent explanation of historical processes: exhaustive because, more than any other idea, the digital revolution has allowed us to read and interpret reality in an immediate and understandable way; coherent because it proposes a clear, stable, and stabilizing vision of the world in which the digital revolution is the dominant paradigm of the present (and of the future), a glue used by contemporary societies to hold themselves together, even to recognize each other, as well as to adopt common habits and dreams.

In this respect, the digital revolution even seems to have an all-encompassing quality because it provides an explanation of the world that applies to every field of human knowledge and practice and, unlike more limited ideologies, allows a global reading of *all* human phenomena. The digital revolution has these characteristics: political parties and politics have changed due to the advent of digitization, the economy has been disrupted by telematics, our ways of communicating and being in the world are not comparable to what they were in the pre-digital era. Think of any field of human activity and you'll find a book, a guru, or a company ready to tell you that the digital revolution has changed it forever.

What's more, this world view has a hold on reality; it guides and transforms societies. It's an idea that influences behaviors, habits, and everyday life. Billions of people take for granted that they are part of an epochal revolution which has allowed them to communicate and *do* things that were impossible or even unthinkable a few decades earlier. So they buy, work, and play—spending most of their time fiddling with digital objects, but above all they have married (and loved) the idea of being immersed in the digital revolution. If, until a few decades ago, the adepts of the digital revolution constituted a niche of technology experts, seen as eccentric people and given odd monikers like *hackers*, *nerds*, or *geeks* (which refers precisely to the semantics of the bizarre), today, billions of people go into ecstasies over the market's latest product launch. The base of the revolution's evangelists has thus expanded to embrace ordinary people, fascinated by the design of digital media, the connective potential of social media, or life-enhancing technologies. In the words of Antonio Gramsci, the digital revolution is a form of *cultural hegemony* that has imposed a benchmark cultural universe on societies, one that we've all internalized and made our own. And in this respect, the digital revolution is a hegemonic ideology because it has in fact colonized much of contemporary discourse, thoughts, and dreams. Note: I don't want to apply any negative associations to the concepts of all-encompassing or hegemonic ideology; it's only a first step in reconstructing, in the most neutral way possible, the historical-cultural path of the digital revolution so far.

However, as mentioned, the word ideology also has a second meaning that refers to a semantics of a *false or distorted way of thinking*. In this sense, the digital revolution could be interpreted as a sort of scam or conscious deception, useful to achieve certain ends and objectives generally of an economic nature. The digital revolution is undoubtedly driven by various partisan interests, led by human beings, but also by companies, objects, and places that, while not representing themselves as a specific political party or group as some of the "classic" ideologies do, have become dominant classes or symbols nevertheless. This second meaning rarely finds space in the narrative of the digital revolution and I personally believe it is misleading. While I argue for the relevance of partisan interests, particularly in the conclusion of the book, I do not believe that the digital revolution is a deception perpetrated by stakeholders who don't actually believe in the revolution itself. Quite the opposite: those protagonists of the revolution who will find ample space in this book *genuinely* believe or believed that the digital revolution would change (or can still change) humanity. Clearly, these people, just like any revolutionary characters and heroes—is there a revolution without its heroes?—have something to gain, but even for the digital revolution's protagonists, the aspect of self-conviction or self-deception seems more decisive and significant than a deliberate deception.

In fact, the ideological vision of the digital revolution is so internalized at the individual and collective levels that it has become a universal thought pattern. We genuinely believe (and believed) that the digital revolution will save us, that it is the best way to enrich ourselves, to feel better, to live in better societies, to communicate more efficiently. Whether it is or not—think for example of how inefficient our communications on social media are and of all of the digital "noise" with which we distract ourselves and that we produce—is only relatively significant. What is significant is the fact that we've made this ideology our own as if it were a "natural," indisputable, rational, and logical idea. There are no alternatives to the digital revolution because, we all believe, the world inevitably moves from an analog past to a digital future. Consequently, many present (and future) revolutions depend on the digital one: the contemporary political, technological, and sociocultural revolutions, as will be highlighted in Chapter 2, all seem to depend on or be intertwined with the digital one, which acts as a trigger for these changes—often considered epochal.

This is also why the digital revolution is a sort of contemporary mythology (after all, ideologies have a mythical component), sometimes full of prejudices: that digital is better than analog, that the transition to digitization improves society from various points of view, that transformation is primarily beneficial. But even when there *are* critical approaches to digitization that underline the growing digital gaps between generations and cultures, or the irruption of widespread and remunerative surveillance, the fact that a digital revolution is underway or has already been accomplished is not questioned. In short, even critical approaches to the digital are ideological with regard to how "revolutionary" it is.

Figure 1 Is every new digital technology a breakthrough, a revolution?
© CartoonStock.com, cartoonist Sunill Agarwal and Ian Baker, uploaded April 6, 2023.

This book then is a short history of the ways in which the digital revolution has been narrated over time, by whom, and with what recurring mantras. Chapter 1 aims to provide a definition, or better, several definitions, of the term digital revolution. There are very few definitions which are widely accepted, with the exception of the famous transition from atoms to bits, especially as theorized by Nicholas Negroponte. Uncertainty prevails, starting with names and periodization. The digital revolution has been given different names over time; the most famous ones are "the information revolution" (especially in the 1960s and 1970s), "the digital revolution" (especially in the 1990s and 2000s), and "the digital transformation" (which has become very popular since the 2010s). The beginnings and the endings of the revolution are also uncertain and, according to some "prophets," the true digital revolution is yet to come. I claim that this uncertainty, rather than being problematic, is part of a crucial strategy to keep the idea attractive and sexy.

Chapter 2 deals with how parallels are made with past revolutions, as well as the idea that the digital revolution is the revolution around which all the present changes revolve. The change wrought by the digital revolution has been compared to that of fire or the book as well as the French or Russian Revolutions, but there is a particularly constant reference to the Industrial Revolution. These parallelisms are crucial to placing the digital revolution "on the shoulders of giants" and to certify its relevance on the basis of this pedigree. However, the digital revolution is also the revolution of our times, and all of the other social changes are inevitably influenced by it: think of political uprisings like the Arab Springs (sometimes renamed Twitter or Facebook Revolutions), the green revolution, the fight for climate change, in which digital transformation seems destined to play a key role, and finally the COVID-19 pandemic, when the digital revolution literally saved some people's lives.

The digital revolution has lasted for several decades and it promises to continue in the near future. This dimension of persistence and permanence is a "classic" characteristic of all the revolutions in the past, as stated in Chapter 3, but it is particularly relevant for the digital revolution because lasting means attracting funds, interest, and maintaining relevance over extended periods of time, even if the same discourse is continually repeated. I have called these repetitive narratives mantras: the disruptive character of the revolution (after the digital revolution, nothing will be the same as before); the fact that it involves the totality of human beings and spheres of action; the fact that it is irresistible and, indeed, that opposing it is counterproductive; the fact that the revolution is never completed, but needs to be continually happening (the future is the favorite tense of the digital revolution); and finally, as Trotsky and others said of the Russian Revolution, it is a permanent phenomenon which has lasted for decades and will last into the future. All of these mantras are buzzwords and tropes of the digital revolution which have been assigned to it without much reflection. Deconstructing their origins and the reasons why they persist over time is the ultimate goal of this chapter.

Chapter 4 aims to understand why people believe in the digital revolution almost as if it were a contemporary religion. In fact, it is a quasi-religion because it is often described using sacred and religious metaphors: the protagonists of the digital revolution are gurus, messiahs, and tireless evangelists; the places where the revolution is happening are the meccas towards which we should turn; digital objects are sacred relics and transport us to transcendent realities. And there are even heretics and infidels who have taken the liberty of opposing the rhetoric surrounding digitization's positive effects but who very rarely question the revolutionary nature of the process.

The Conclusion aims to answer two simple questions: who profits from the grand narrative of the digital revolution? And why does it keep going (or seem to continue with little debate)?

All these chapters focus on the development of specific discourses over time, starting after the Second World War but often even earlier. Nevertheless, the book doesn't provide a complete survey of all the possible examples of the revolutionary discourses, or of all the fields that have been revolutionized by the digital revolution. That would have been both an impossible and an uninteresting task to accomplish. But you, the reader, can play a game: try to find other fields, other discourses, other technologies or gurus or places which are not mentioned here and apply the theoretical background of this book to them. If the persistent discourses which are continually applied to the digital revolution can be applied to other examples which are not included here, the book itself gains more value. It can be considered a universal key, a *passepartout*, for interpreting the revolutionary character of the digital revolution and could even inspire new books, papers, or research projects that aim to expand on this framework. This is what I hope for both the book and its readers.

Endnotes

Regarding the history, the semantics, and the concepts of ideology there is a huge amount of literature. I have found particularly useful for this book Cormack, 1992; McLellan, 1995; Rossi-Landi, 2005, and Rehmann, 2007.

Antonio Gramsci (2011) deals with the topic of cultural hegemony in a number of his works, but above all in the three-volume set of his *Prison Notebooks*.

Regarding the link between ideology and technology, a number of works are cited in the book, but for a broader framing of the subject see David Nye (1990, 1994), who links an analysis of rhetorical/ideological responses to radical technological change with an in-depth exploration of how that change actually manifested itself. Also Leo Marx classified the concept of technology into two categories: ideological and substantive. He then quotes a speech delivered by US Senator Daniel Webster at the opening ceremony of a new railroad in 1847, which is full of the religious ideology of revolution which we will try to retrace in this book in the context of the digital revolution: "It is an extraordinary era in which we live. It is altogether new. The world has seen nothing like it before. I will not pretend, no one can pretend, to discern the end; but everybody knows that the age is remarkable for scientific research into the heavens, the earth, and what is beneath the earth; and perhaps more remarkable still for the application of this scientific research to the pursuits of life. The ancients saw nothing like it. The moderns have seen nothing like it till the present generation.... We see the ocean navigated and the solid land traversed by steam power, and intelligence communicated by electricity. Truly this is almost a miraculous era. What is before us no one can say, what is upon us no one can hardly realize. The progress of the age has almost outstripped human belief; the future is known only to Omniscience" (Marx, 2010, p. 564).

When I started writing this book, I would have expected more literature directly connecting ideology and digitization. I have found useful De Biase, 2003; Bowers, 2014; Kroker & Kroker, 1999; Mounier, 2018; Zacher, 2015 (especially the conclusion *Les humanités aux prises avec l'idéologie numérique*).

Much more common are the studies that highlight forms of ideology linked to single digital technologies and, in particular, the Internet. To cite some examples: Musso, 2003 (partially translated in Musso, 2017, which offers a history of the concept of the network and even discusses "retiology," a form of idolization of this concept); Bory, 2020; Fuchs, 2020; Sarikakis & Thussu, 2006. Christian Fuchs is perhaps the author who has made the most in-depth study of some of the aspects and ideological consequences of contemporary digital technology.

Regarding the need for studies that analyze the digital revolution from a historical perspective, see Agar, 2019, p. 2; Ensmenger, 2012; and Köstlbauer, 2011.

1

Defining the Revolution
Blessed Uncertainty

Many people are convinced that they're living at a time when the digital revolution is an established fact, but there are very few clear definitions of this revolution. Which definitions have been imposed over the decades and which ones have been discarded? In what historical periods did they come into use and when were they most popular? Finally, what different names have been applied to the digital revolution over time? These three questions will serve as a guide to this chapter, which aims to reconstruct the main ways in which the digital revolution has been characterized, periodized, and renamed, pointing out ambiguities and uncertainties which, contrary to what one might think, give the revolution strength and vitality.

1.1 Transform and Transform Again: From Atoms to Bits

There are at least two tools that help us understand the evolution of concepts, and obviously not just the concept of the digital revolution: Factiva and Google Ngram Viewer. Factiva is a database owned by Dow Jones & Company that collects content in 28 languages from more than 30,000 digitized sources including newspapers, magazines, corporate documents, television and radio broadcast transcripts, and images as well as many other sources. Google Ngram is a freely accessible online search engine that allows you to track the frequency of certain words in the millions of books that Google has digitized over time.

By inserting the string *digital revolution* in both databases, starting from 1960 up until the present day, we obtain two graphs and various information useful to understanding how the term has evolved. Specifically, the expression "digital revolution" has been used especially since the 1990s; it experienced a first peak of popularity between the end of the 1990s and the early 2000s and a second around 2017–2018.

Factiva allows you to associate the most frequently cited companies, institutions, and personalities with this search term. Among the former are the European Union, Alphabet (and therefore the entire Google galaxy), the BBC, Microsoft, the United Nations, Liberty Global, Apple, along with a mix of public and private institutions which have dealt with digitization policies and strategies over time. Leading players include Indian Prime Minister Narendra Modi, and one of his country's best-known entrepreneurs, Mukesh Dhirubhai Ambani, as well as Donald Trump, Steve Jobs, Bill Gates, Rupert Murdoch, and Mark Zuckerberg, among others.

This data provides a general and quantitative view of how often the expression *digital revolution* and related terms are used; however, it doesn't indicate if there is a positive or negative vision of the digital revolution, in which contexts it is named (ironic or serious, for example), or even if it appears in more or less prestigious magazines. However, this data can be useful for understanding the broader structure of the digital revolution, its popularity over time, its association with powerful digital companies and institutions or "mythical" entrepreneurs, as well as with politicians or specific geographical locations (evidently, areas of the world such as the United States—the real driving force of the digital revolution, at least at the beginning—India, and Europe stand out in these initial observations).

But what do we mean when we talk about the digital revolution? Are we sure we can converge on an unambiguous definition which captures the primary elements and is easily understandable? As very often happens with popular concepts and ideas, even the characteristics of the digital revolution are not well defined. Its qualities are taken for granted, implied, or, once again, stereotyped formulas are obsessively repeated. A symbolic starting place would then be Wikipedia: the entries of the most famous online encyclopedia are also very valuable historical sources for understanding how common discussions and ideas settle over time. In the English version from early 2023, the digital revolution ("also known as the Third Industrial Revolution") is defined as:

> the shift from mechanical and analogue electronic technology to digital electronics which began anywhere from the late 1950s to the late 1970s with the adoption and proliferation of digital computers and digital record-keeping, that continues to the present day. Implicitly, the term also refers to the sweeping changes brought about by digital computing and communication technologies during (and after) the latter half of the 20th century. Analogous to the Agricultural Revolution and the Industrial Revolution, the Digital Revolution marked the beginning of the Information Age. Central to this revolution is the mass production and widespread use of digital logic, MOSFETs (MOS transistors), integrated circuit (IC) chips, and their derived technologies, including computers, microprocessors, digital cellular phones, and the Internet. These technological innovations have transformed traditional production and business techniques.

This is a very layered definition and contains a number of elements which will be discussed in this book. First, the digital revolution is identified here as a technological transition from an analog paradigm to a digital one, two universes which are considered incompatible but in fact are not. This definition has been generally accepted for quite some time but it was Nicholas Negroponte who popularized it in his 1995 bestseller *Being Digital*. Although the term "digital revolution" appears in his book only four times, Negroponte theorized the transition from a physical world made up of atoms to an immaterial one made up of bits. In his opinion, this radical and paradigmatic shift would lead to a new world, unrecognizable when compared to the past.

There is a concrete example of this idea. In an extensive *National Geographic* article dedicated to the information revolution, also published in 1995, there is a photo of Bill Gates in a climbing harness, hoisted atop a towering stack of 330,000 sheets of paper and holding a CD-ROM. All of this to illustrate that "[t]his CD-ROM can hold more information than all the paper that's here below me." It's a symbolic representation of the transition between atoms and bits theorized by Negroponte, moreover promoted by one of the gurus of the digital revolution.

A second characteristic of the digital revolution that emerges from the Wikipedia entry is the fact that it is a transformative revolution for human societies, capable of unleashing social, economic, and political changes that have led (or will lead) to a new, information-centered society. In other words, the social changes of the last few decades are often described or narrated as a mere *consequence* of the digital revolution. This conviction goes beyond classical technological determinism, according to which technologies are the cause of linear social change. It is an abstract entity like the digital revolution which (also thanks to the new technologies that it constantly churns out) becomes the engine of transformation and therefore the primary cause or agent of change.

The transition from an analog paradigm, concrete and composed of atoms, to a non-material, digital one, together with the radical changes that have affected contemporary societies (the Chinese version of Wikipedia argues that it is "the largest and most far-reaching scientific and technological revolution in human history, and it has not yet ended"), are the two common denominators of the best-known and most accepted definition of the phrase "digital revolution." Both are debatable, but these two narratives have been the most relevant ideological pillars that support our mental image of the digital revolution. But are these two characteristics sufficient to define it? Two pillars alone are not enough to support a stable construction.

In fact, the Wikipedia definition introduces other open questions that will be examined in this book. For example, the temporality of the digital revolution is quite a controversial issue. According to the Italian

"I JUST INVENTED THE 'ONE' AND THE 'ZERO'.
LET THE DIGITAL REVOLUTION BEGIN!"

Figure 2 The digital revolution written in stone.
© Glasbergen. Reproduced with permission of Glasbergen Cartoon Service.
Cartoonist Randy Glasbergen, Unknown year.

version of Wikipedia, it starts "during the late 1950s," yet in the English and Spanish articles the subsection on "Origins" indicates a time frame from 1947 to 1969, while in the sophisticated French entry, which highlights the ambiguity and ideology of the concept, its genesis dates back to the first automatic and computing machines of the eighteenth and nineteenth centuries, and in the German version it refers to the "end of the twentieth century." In Section 1.2 of this chapter we will explore the periodization of the digital revolution and, above all, why it is so variable.

Another question, touched upon not only in the English version of the definition, concerns the name: the digital revolution has been referred to in multiple ways, using terms that may seem synonymous but which actually hide different interpretations of its transformative power. Of course, these terms are also the natural result of the historical moment in which they were coined. Section 1.3 of this chapter focuses on this topic. The analogy with the Agricultural and Industrial Revolutions (both with capital letters) is another way to narrate the digital revolution, in which it is presented as the grandchild of these relevant turning points in human history, and I will focus on this aspect in Chapter 2. A final aspect mentioned in the English Wikipedia entry is the link between the digital revolution and new information and communication technologies (ICT); in particular transistors, integrated circuit chips, computers, digital cellular phones (up to smartphones), as well as the Internet. Many other technologies that, over the years, have been seen as disruptive and revolutionary could be added to this list; some of them will be defined as relics in Chapter 4.

1.2 Dating and Re-dating: Variations in Periodization

Can we assign a definitive date to the digital revolution? Is there consensus on when it originated and when it will eventually end? These are not easy questions to answer.

The "official" accounts often view cybernetics not as the beginning of the digital revolution but rather as its main source of inspiration, a prelude to and a trigger for subsequent transformations. The expression *cybernetics* was introduced in 1947—the same year that the transistor, often considered a key object of the revolution, was invented—and designated a new science focused on the interaction between computers (and more generally machines) and human beings as well as the phenomena of self-regulating communication systems. According to US mathematician Norbert Wiener—one of the patriarchs of the revolution, as will be pointed out in Chapter 4—these two aspects would be able to generate the most relevant social transformations in the future.

Beyond the guiding light of this discipline, which moreover may not represent the trigger but rather one of the symptoms of the revolutionary phenomenon, the periodization of the digital revolution varies widely. As mentioned, some believe it has roots in the eighteenth and nineteenth centuries, but most accounts focus on the twentieth century. The 1940s and 1950s are sometimes referred to as the beginning of the digital revolution because that was when the first large computers, known as *mainframes*, initiated a computing revolution. Others believe that the 1960s was a watershed decade because it was the period in which the Arpanet network, now recognized as the early foundation of the Internet, was installed. Some scholars define the 1970s as *really* crucial because the market saw the launch of new products, such as microprocessors and personal computers, and the so-called information revolution started to become one of the most representative metaphors of contemporary society at the time. The core concept of both post-industrial and information society was that computers and the new telecommunication networks (there was as yet no talk of the Internet) were the fundamental tools necessary to move from an agriculture- and industry-based economy to one based on services and information. Jean-François Lyotard, a postmodernist theorist, also identified the information revolution that was taking place in the late 1970s as one of the keys to understanding the transition from a modern to a postmodern society. In the 1980s, computers were slowly introduced into many homes, especially in Western countries, above all used as a gaming tool: for this reason some scholars identify this decade as marking the *real* explosion of the digital revolution.

However, as the analysis of the term's popularity mentioned at the beginning of this chapter confirms, the most popular and crucial decade for the development of the idea looks like the 1990s. These were the years that saw the widespread use of mobile telephony and the launch of the World Wide Web, still one of the Internet's primary applications. This is the decade in which Bill Clinton and Al Gore, respectively president and vice president of the United States, drew up the plan known as *The National Information Infrastructure: Agenda for Action* (1993), better known as the "information superhighway" project, which aimed to enhance and digitize the country's telecommunications networks and which was then copied and adapted in many other countries. The first paragraph of this document states: "Development of the NII (national information infrastructure) can help unleash an information revolution that will change forever the way people live, work, and interact with each other." Therefore, the mantra of a *total* revolution (which will be explored more fully in Chapter 3) was already well established. The early 1990s was also when *Wired*, probably the most symbolic and authoritative magazine of the digital revolution, a kind of paper evangelist, was conceived. This publication will be frequently cited in this book as it contributed to spreading, or rather propagandizing, the existence and relevance of the revolution itself.

Yet other observers hold that the digital revolution *really* got started in the 2000s, when first millions and then billions of people began to use their smartphones to access the Web. According to this theory, a revolution is fully realized when its effects and its products are available to the masses, something which has actually only happened in the last few decades, moreover unequally in different regions of the world.

The periodization of the digital revolution is complicated by the fact that in different countries and cultures, digitization has happened at different times, often lagging behind the United States or other digitally advanced Western countries. China, India, and Russia are countries that have been considered "on the verge of" embracing the digital revolution for decades (the term *on the cusp of* is obsessively repeated in international publications). More recently, these same nations have been mentioned as candidates to "lead" this revolution in the world of tomorrow. These claims can be found in the pages of the *China Daily, The Moscow Times*, or *The Times of India*, but it can also be heard in the official speeches made by politicians and entrepreneurs in these countries. While this narrative emerged in China in the early 2000s, in India it is a phenomenon that dates back to 2010s. For example, in 2018 the aforementioned Prime Minister Narendra Modi remarked that India was on the threshold of the Fourth Industrial Revolution (one of the many synonyms of the digital revolution) and that it would lead this revolution in the future. Other media outlets, such as *AllAfrica* or *Gulf Business*, mention a digital revolution capable of creating progress and resulting in a definitive leap into modernity in the most backward areas of Africa, Asia, or Latin America. In short, it seems that the rhetorical force of the digital revolution has shifted—without any substantial reinterpretation—from the United States, where it erupted in the 1990s and early 2000s, to other countries in the world.

The digital revolution has also happened at very different moments in various media sectors and in everyday life. This is another factor that has contributed to complicating any definitive dating of the revolution. To stay with media, for example, the music industry was among the first to be transformed by digitization with the emergence of the CD as early as the 1980s, followed by the MP3 format in the 1990s, and then with music streaming services starting in the early 2000s. Meanwhile, other media sectors were affected by the phenomenon of digitization much later, some—for example radio and book publishing—have been controversial and perhaps will never complete the digitization process.

Even more interesting and vague are the discussions around the supposed end of the digital revolution; after all, revolutions have many beginnings and rarely have an end. Some observers argue that the digital revolution is over because it has dominated and definitively transformed our way of life. This hypothesis, which emerged both in the 1990s and in the 2000s, states that we are already living in a digital universe, or better yet in a digital age, to which the revolution was the prelude. Additional proof is then offered by the thousands of history books that end with a chapter dedicated to the digital age, thus taking for granted that it has already happened and is well established. Another, less popular version of the story says that the digital revolution ended because it actually lost: it failed to keep its promises and therefore has lost its appeal. A third narrative, strongly supported by businesses and digital gurus, believes that the digital revolution is not over and that, of course, the best is yet to come. We will return to the subject later but here it is interesting to note how historians have often focused on the beginnings of revolutions and not on their final phases, because these are the most ambiguous. In the case of the digital revolution both the beginnings and the ends are ambiguous, but they are both continuously commemorated and its traditions are rhetorically reinvented and constructed day by day.

Figure 3 When did the digital revolution begin to end? A tip for future historians.

© and reproduced with permission of the cartoonist Bradford Veley, unknown year.

For all of these reasons, the digital revolution must be referred to in the plural; there is not *one* revolution, but rather many digital revolutions and many phases within these revolutions. Companies like Britain's Ovum announced triumphantly in 2016 that "a fresh wave of digital transformation, the Second Digital Revolution, is coming because of cheap and readily available hardware, processing, and connectivity as well as new, disruptive technologies" (these words, by the way, work perfectly well in any era and for any technology). According to others, such as Christopher Barnatt, the second digital revolution has another meaning: if the first occurred in the 1980s and 1990s and was based on the Negropontian axiom of the transition from a world composed of atoms to one composed of bits, the second can be traced back to the early 2000s and is characterized by an opposing, re-atomization process because, in the meantime, billions of physical devices and digital products have invaded the market. This rhetoric of the new digital revolution, in a paradoxically analog style which heralds a return of the digital to physical reality, has met with some success and has been adopted several times by the gurus of the fab labs (fabrication laboratories), small workshops that offer digital manufacturing services to the public, which have popped up in various countries around the world. As early as 2006, Neil Gershenfeld, director of MIT's Center for Bits and Atoms in Boston, supported the advent of a *third digital revolution*, following the communications and computing revolutions, which was centered on the ways in which digital objects were manufactured.

Naturally, we can also talk of a fourth or a fifth digital revolution, but instead of pursuing these or other periodizations, which are destined to age rapidly, a central feature must be recognized: the main temporal characteristic of the digital revolution is precisely its flexibility. And, what's more, continuous restatement is an intentional marketing strategy imposed by the reality of the facts. As early as 1988, Victor Scardigli pointed out how the digital revolution progressed in waves of illusion and disillusion. In fact, over the decades this ideology has gone through various moments of crisis: among the most symbolic, the bursting of the *dot.com* bubble in 2000–2001, the 2013 revelations of former CIA and National Security Agency (NSA) analyst Edward Snowden about the mass digital surveillance programs that the United States implemented in various countries around the world, and probably the financial and social crisis of digital platforms that emerged after the COVID-19 pandemic. Yet even in these critical moments, when the vision of digitization as an agent of radical societal transformation could have declined, sacrificed on the altar of economic failures or mass surveillance, the conviction of living in a digital revolution (or related expressions) persisted. And it was precisely this continuous restatement that played a fundamental role in keeping the digital revolution ideology alive, alongside its ability to recycle or rather repropose itself in various markets. Moreover, the fact that the global brand of the revolution has been adopted at different times in different regions of the world means that it acquires both a global and local temporality, leaving it ample room to reach, as if for the first time, even those territories where it has not yet managed to break through. So, it's possible that over the next few years a new revitalization of the revolution, even from a retrospective view, will emerge as a consequence of new crises which might develop. After all, the present always influences the past and the new needs of the digital revolution (or transformation) will also guide its history.

1.3 Baptizing and Rebaptizing: Parallel Expressions

Even the names by which the digital revolution has been called over time are extremely variable, and, like the one chosen as the title of this book ("digital revolution"), they should actually be interpreted as a cloud of expressions that are partly equivalent and constantly changing. Not all these names or labels can be considered synonymous, but the historian Reinhart Koselleck recalls how the history of concepts must "register the variety of names for (identical?) materialities in order to be able to show how concepts are formed." In short, concepts cannot be studied by only examining the best-known term that designates them—in our case "digital revolution"—but should also take into consideration "parallel expressions." And there are at least six different families of expressions parallel to digital revolution which have preceded and accompanied it over the decades and which will perhaps replace it in the future.

The first family focuses on the revolutions brought about by a single technology. The most popular terms in English, the "official" language of the revolution, are computer revolution, Internet revolution, and mobile revolution. All three expressions have distinct histories and temporalities. The computer revolution precedes the other two and this term has been used at least since the 1960s, the Internet has been seen as the revolutionary technology par excellence since the 1990s, and the mobile phone (and now increasingly the smartphone) since the early 2000s. There are also differences in geographical terms: for example, the expression *mobile revolution* has long been in vogue in regions of the world such as Africa, Latin America, and parts of Asia, which have not experienced the wave of the "first" digital revolution, and in which mobile telephony has not been preceded by dozens of other digital devices (all invariably revolutionary). The computer, Internet, and mobile revolutions have often been hybridized: once again the most cited example is the smartphone, seen as the flagship product of this convergence between revolutions and which, moreover, is both a computer and a mobile phone with access to the Internet. Finally, the terms for these three revolutions also include variants, such as the cyberspace revolution or the Web revolution for the Internet, or the smartphone revolution for mobile. This is interesting for at least two reasons. Over time, if substantial innovations are introduced, the same technology can be revolutionary more than once. Moreover, semantic flexibility is necessary to remedy the rapid obsolescence of revolutionary appeal: one cannot state for decade after decade that the computer, the Internet, and mobile devices are disruptive because this form of narration tends to age rapidly. There is a need to continually renew the conversation about revolutionary significance, to adapt to techno-social changes, and to assign new labels to old concepts or technologies.

The first and longest-running synonym for digital revolution tied to a single technology is *computer revolution*. As already mentioned, since the Second World War, the computer has been considered *the* tool capable of profoundly changing contemporary societies and of giving rise to what are referred to as post-industrial societies and information societies. The computer revolution, like the digital one, goes through various technological phases; following is a brief list of the most well-known: the big mainframes in the 1950s and 1960s, personal computers between the 1970s and 1980s, computers able to access the Internet starting from the second half of the 1990s, portable devices since the early 2000s, ubiquitous and wearable computers today, and probably quantum computers tomorrow. In short, even the computer revolution has been renewed and reproposed over time with new labels that differ from the previous ones.

A second family of expressions, similar to the technological ones, includes all of those referring to materials. Especially in the 1970s and 1980s, the microelectronic revolution became a popular expression. This term was based on the invention of the transistor through to integrated circuits (also called chips, as mentioned above). Microelectronics made it possible to implement a miniaturization process in industrial electronics and computer electronics in particular. So, the digital revolution is also a revolution of size and, specifically, the shrinking of technological devices. The silicon revolution, another example from this family of expressions, is often used as a synonym for a digital revolution—with a vague implication of synecdoche, meaning a part is made to represent the whole—and derives from the use of this chemical compound in networks and digital devices that have become part of our daily lives. The longevity of this expression also lies in the fact that it was adopted as a "geographical" brand and, even today, one of the most famous meccas of the digital revolution is Silicon Valley.

The third family of parallel expressions for the digital revolution highlights how new information technologies remix entire sectors or categories of previous thought. In particular in the 1970s and 1980s, numerous neologisms were coined to express the convergence of a number of previously separate sectors and the birth of new paradigms: think of technetronic (from the crasis of technology and electronics), to telematics (telecommunications and informatics), or the least used "compunication" (from computers and communication). The latter term is contained in the well-known 1978 report drawn up by Simon Nora and Alain Minc and addressed to the French government. Reflecting on the crisis of Western democracies—another topic that is always in vogue—the two senior French officials stated that *compunication* would represent a "Copernican revolution" with "unimaginable" consequences for politics, the economy, and society for decades to come. It's surprising how the consequences of the digital revolution are still considered disruptive today; these consequences always involve and substantially transform the same sectors and will inevitably take place in the near future. Only the words are updated: compunication, for example, has not only disappeared from the jargon of digitization but it also sounds extravagant and has been replaced by other terms that, within a few decades, will probably seem equally bizarre.

The next family of parallel expressions, on which I will dwell further, sees the digital revolution as one of the contemporary Industrial Revolutions: the third, if one follows the English Wikipedia entry quoted at the beginning of this chapter, or the fourth, as expressed in the contemporary buzz-tag "the Fourth Industrial Revolution," used since at least the mid-2010s. Even if the genealogy is not completely clear, it's generally held that the First Industrial Revolution centered on the invention of the steam engine in the eighteenth century, the second on electricity in the second half of the nineteenth century, the third on computers and information technology in the second half of the twentieth century, and the fourth was linked to automation, artificial intelligence, and "industry 4.0" at the beginning of the twentieth-first century. The digital revolution is considered the cause, as well as an element, of the last two Industrial Revolutions, in a play of references that makes it indistinguishable from them, as noted by Baidu Baike, the real Chinese Wikipedia.

A fifth family of synonyms for the term digital revolution combines the information revolution and the communication revolution, two popular but at the same time ambiguous expressions also used jointly when referred to as the ICT revolution (the information and communication technologies revolution). Used especially since the 1990s, this label emphasizes the acceleration in the exchange, distribution, and consumption of information brought on by the digital and its technologies. There are many information or communication revolutions and, for the purposes of this book, it's interesting to consider which ones have been associated with the digital revolution. Once again, opinions are mixed. Some authors, such as Daniel Headrick, believe that the modern information age started in the eighteenth century and that the current acceleration (Headrick uses the terms "acceleration" and "revolution" synonymously) is in continuity with some phenomena from more than three centuries ago. James Beniger finds the origins of information society in the nineteenth century, with the birth of telecommunications as well as modern transport and distribution systems. Gerald W. Brock believes that the first information revolution began with the telegraph and ended with the radio, and that the second, which began in the 1940s with the transistor, radar, and the first digital computers, is still ongoing. On the contrary, there are authors who see contemporary information society as an era distinct from the past, and most of them set its beginning in the 1960s and 1970s.

As already mentioned, the computer is the watershed tool in this definition of the information revolution, and global enthusiasm and hopes are gathered around this technology. For example, as early as the 1980s, there is a statement that "[p]ublicity about the 'information society,' 'information revolution,' 'computer society' and related concepts, has been inundating the media in China," a country described as "an intensely 'computer positive' society: one which accepts the information revolution as a good thing as well as a necessity." Starting in the 1990s and up until today, the terms "information and communication revolution" and "digital revolution" have begun to be used synonymously.

The 1990s are also the period in which, according to the afore-mentioned Factiva and Google Ngram databases, the term *information revolution* experienced its peak popularity. The period of most intense usage is at the end of the decade and the term's decline begins in the aftermath of the bursting of the speculative dot.com stock-market bubble in 2001. The top company (by number of occurrences) associated with the research on Factiva appears to be SoftBank Group Corporation, a Tokyo-based financial holding company that offers digital communication services, founded in the early 1980s by the Japanese entrepreneur Masayoshi Son. Other companies associated with the information revolution include Microsoft, Alphabet, Intel, and, among the personalities, the previously mentioned Mr. Son, Bill Gates, Steve Jobs, Donald Trump, and Xi Jinping. All in all, these results are close to those obtained with research relating to the digital revolution and which therefore indicate a certain overlap between the two concepts. However, the information or communication revolution would seem to be a more flexible brand, more adaptable to different technologies over time, and therefore potentially more lasting and less subject to aging than the term digital revolution. So it comes as no surprise that *information revolution* has been reused for very different technologies, essentially repeating the same themes: among these there are also some non-digital tools, such as magnetic tape, disks, telex, new types of telephone services, new forms of radio introduced in the 1970s, or some digital technologies now considered "old" (even digital is aging rapidly), such as satellite dishes or CD-ROM.

You will have noticed that all of the "parallel expressions" to *the digital revolution* mentioned so far are missing the adjective "digital," a true keyword in the last several decades around which a number of investment streams have been generated. In fact, the sixth and final family of parallel expressions puts the emphasis on *digital* with phrases such as *digital disruption*, *digital transition*, and, above all, *digital transformation*. Digital disruption adds an even more disruptive "quality" to the dimension of the digital revolution: digital is presented as an interruption, a profound or a violent break with respect to the past. Digital transition is the most recent expression; it focuses on the idea of a passage and a move from one state to another, and it's often combined with the idea of a series of transitions (i.e., of energy) which are affecting and will affect human beings. Digital transformation goes in the opposite direction: the expression refers to a harmonic process of change, in some ways broader than the revolutionary one and capable of generating a new form (transforming is composed of *trans-* in the sense of "beyond," and to form or "give shape") and even changing the nature of human society. This dimension also emerges in the Wikipedia pages dedicated to the concept of digital transformation, which highlight the organic and harmonious elements that underpin this idea: in fact, *digital transformation* affects "all aspects of human society" (Spanish), it acts "in an organic and combined way" and "pervasively creates new connections between people, places and things" (Italian), it is "an ongoing, far-reaching process of change in business and society that has been triggered by the emergence of increasingly powerful digital techniques and technologies" (German). The *digital transformation* label, which has joined and may potentially replace *digital revolution*, seems to smooth out the conflicts generated by a revolution, referring to a soft and peaceful transition, and therefore not as violent or brutal as the concept of a revolution. Indeed, one could provocatively argue that the digital transformation is more analog than the digital revolution because the latter foresees only two states: the first analog and the following digital. But the COVID-19 pandemic may have contributed to modifying the meanings of digital transformation and, paradoxically, bringing them back toward the revolutionary dimension: for example, the previously mentioned *digital transition* or the expression *acceleration of digital transformation* have begun to make their way into popular use, making reference to a more frenetic and less peaceful type of change. At the virtual G20 meeting held in early 2021, attendees were reminded, among other things, that "lifestyle and consumption patterns changes foisted by the pandemic restrictive measures have led to an unprecedented acceleration of the use of new technologies, which many observers tend to define as a digital revolution." It would seem that even a pandemic can help reinvigorate the myth of the digital revolution and its related expressions.

The growing popularity of "digital transformation" emerges in both Factiva and Google Ngram and the term is now used more frequently than digital revolution. Microsoft is still among the most cited companies, however it is now followed by the Chinese Huawei, the French conglomerate Schneider Electric, and India's Infosys Technologies Limited, an IT services company based in Bangalore. Among the most frequently mentioned people are Microsoft CEO Satya Narayana Nadella, born in India and now a naturalized US citizen, along with politicians like Vladimir Putin, Joe Biden, Xi Jinping, Narendra Modi, and Donald Trump, who had already emerged as relevant figures in all the parallel expressions for the digital revolution. Here a second substantial difference emerges between the digital revolution and the digital transformation: the language of the transformation tends to be less and less American English and more and more Chinese and Indian (or rather, the myriad of dialects spoken in India). If the digital revolution was and is American, the transformation appears to be Asian.

And yet, *digital revolution* and *digital transformation* are phrases often used almost interchangeably. In particular, as the English Wikipedia entries remind us, digital transformation and the digital revolution both involve the adoption of digital technologies in order to convert analog information and systems through the use of computing and communication technologies. Furthermore, the digital transformation seems to be linked to the future as the digital revolution has been: forecasts by specialized companies such as Gartner paint a bright horizon for digital transformation. They remind us of how many regions of the world have yet to be reached by the panacea of digitization (without doubting that it will happen, it's just a matter of time), they take care to point out that the complete digital transformation is still far from being achieved. Finally, virtually every government in the world has adopted or is adopting a strategy and policies to address digital transformation. These strategies are sometimes limited to renaming old ministries in an attempt to spruce up dusty public offices with a veneer of innovation. In 2001 Italy established the Ministry for Technological Innovation and Digital Transition, in Ukraine the ministry of Digital Transformation was launched in 2019, in Spain it was the Ministry of Economic Affairs and Digital Transformation in 2020, and in Mauritania the Ministry of Digital Transformation, Innovation and Administration Modernization in 2021, and the list could go on. It is probably more interesting to note that, from a political point of view, huge global investments are consistently converging around the key term "digital transformation," just as they have converged, in the recent past, around the idea of digital revolution.

Are all these ways in which the digital revolution has been named and renamed therefore equivalent? Do they express similar concepts and ideas or are they even interchangeable? On the one hand, in different historical periods, all the parallel expressions mentioned have had, and in some cases still have, the function of signaling that a social change caused by information, by digitalization, or by some technologies or materials was taking place. On the other hand, however, these "brands" are not entirely equivalent. They have been adopted in different historical periods and so can be seen as daughters and sons of sensitivities and societies that have changed over time. They are terms designed for technologies that are profoundly different from each other: telex, microelectronics, the Internet, smartphones, and quantum computing have all been seen as revolutionary technologies capable of subverting the social order but they cannot be considered comparable. Finally, as mentioned, in some of these expressions the digital dimension is lacking or not yet central. In this aspect the information revolution is representative, because in the 1960s and 1970s it was indeed associated with computers, which at the time were not yet fully digital tools, but also

Figure 4 Defining "digital transformation" is not easy, right?

with many technologies that we would now define as analog. On the other hand, as is already argued by some scholars, it is necessary to look at historical moments in which the terms digitization and digital were not yet well defined in order to fully understand the digital revolution. It can perhaps be concluded that the different names with which the digital revolution has been baptized and renamed (what, paraphrasing Koselleck, we have called parallel expressions) have contributed to forming and popularizing the idea of a revolution brought about by digitization and, at the same time, they have been absorbed by and integrated into this same idea, like a river in full flood that, in recent decades, has overwhelmed all the conceptual sources and roots that generated it or that flanked its flow.

1.4 New Wine in Old Bottles: The Secret of Eternal Youth

This chapter has shown how the digital revolution is a lasting ideology which came into being no later than the end of the Second World War, or perhaps much earlier. An ideology in constant flux over the course of its history, which has been enriched with definitions, periodizations, and new names but also "revolutionary" objects and people. In short, contrary to popular belief, the digital revolution has been affecting contemporary societies for quite some time, despite the fact that it is always discussed as a recent phenomenon, which has just emerged, which is changing humanity, and which will change it even more in the future. Why? I believe that the secret of the revolution's eternal youth, its "blessing," lies precisely in its instability and ability to renew itself. Definitions, dates, and even names of the digital revolution are uncertain or, better still, adaptable and flexible. The digital revolution is like a chameleon, it knows how to adapt to the uses and fads of the moment and adopt new relationships with time and brands, but deep down it is always the same. Indeed, it could be said that this "uncertainty" or "steerability" of the digital revolution is a survival strategy: to remain seductive and not lose its attractiveness, the revolution must necessarily be renewed. The new periodizations that propose a first, second, third, and fourth digital revolution (and in the future the saga will continue), like the new names—above all "digital transformation"—are a necessity. Like a snake sheds its skin when the outer layer is no longer flexible, so the digital revolution must change some element when its thrust seems to wane, be aging, or is just losing its attraction. However, the change only occurs in the most superficial layer (the sexiest name or the latest and smartest device replaces the previous ones or again a new "big event" or "next big thing" that forces the redating of previous histories), while the deeper layer, the underlying arguments and narratives, the true ideological core, always seems to remain immutable.

It is the classic mechanism of "new and always the same" that the digital revolution has made its own. This mechanism, defined by Edgar Morin in the mid-twentieth century as one of the crucial characteristics of mass culture, has also been used by the digital revolution and all of its narrators. And it is also an element that can be traced back to the very etymology of the word *revolution* which, as both Hannah Arendt and Reinhart Koselleck, two cornerstones of revolutionary thought, remind us, is at least dual. One meaning, predominant at least until the French Revolution, claims that revolutions are cyclical and recurrent movements. The term is used especially in astronomy to indicate that stars and celestial bodies return to their starting points after having completed an orbital revolution in a certain period of time (after all, Copernicus wrote *De revolutionibus orbium coelestium*, the title commonly translated as *On the Revolutions of the Heavenly Spheres*). Nothing would seem further from the other meaning of revolution, the one most used today, which instead indicates an event capable of breaking with the previous cycle, of subverting an old order and creating a new one. The digital revolution seems to be able to combine both meanings of revolution: on the one hand, it is certainly narrated as a radical break from the old order and the analog world, but on the other it also provides for continuous movement, with some aspects of regularity, return, and the re-emergence of topics that are new and always the same (including its disruptive nature).

Endnotes

As stated in Section 1.1, the definition of "digital revolution" is taken from the English Wikipedia entry dedicated to the topic. The text also refers to the Italian, French, Spanish, German, and Chinese Wikipedia entries. I consulted these web pages, as well as the others mentioned in the book, in December 2022; however, like all Wikipedia entries, the one on the digital revolution is unstable and can change over time. For example, the first version of the English-language entry for "Digital

Revolution" (with capital letters) appeared on Wikipedia in 2007, but the first readable version dated back to March 2009: "The Digital Revolution (1980–2010) is the change from analog electronic technology to digital technology, that has taken place since c. 1980 and continues to the present day. Central to this revolution is the transistor and its derived technologies, including the computer, cellular phone, and fax machine." This first version is significantly different from the current ones, in terms of periodization and technologies which have involved, including even the fax machine which would be placed under analog devices today. But there are also several continuities like the shift from analog to digital, and the role of transistors, computers, and mobile phones. This is also why I think it is quite interesting to start from the Wikipedia entries, since different ideas and definitions have settled in over time. Surprisingly, the major international dictionaries, including the *Oxford English Dictionary*, have not dedicated a specific entry to the term "digital revolution." The text also mentions Negroponte, 1995 and the photo by Bill Gates can be found in Swerdlow, 1995.

With regard to the periodization of the digital revolution discussed in Section 1.2, the full text of *The National Information Infrastructure: Agenda for Action* from which the quotation is taken is available online. Among the dozens of articles on the digital revolution in China that appeared in *China Daily*, see Barris, 2014; Keju, 2018; Yu'an, 1994. For Russia, see "The Information Revolution Is Coming" 2003, where Soviet Culture Minister Mikhail Shvydkoy recalled how the digital revolution was a cultural revolution ("Russia a Leader in the Information Revolution", 2013). Modi's statements about the fact that India was destined to lead the digital revolution are from "India at the Cusp of Digital Revolution", 2018. Other interesting articles on India and the digital revolution are "The Epicentre of the Digital Revolution", 2014; and Dass, 2020. On the fact that the digital revolution is over and that, consequently, humanity is already living in a digital age, some similar observations years later can be seen: for example, Davies, 2009, p. 14; Gunnerson, 1993, p. 70. Ovum Ltd's prediction is taken from Hartley,

2016. Also cited are Barnatt, 2001; and Gershenfeld, 2006. On the illusions and disillusions of digital, see Scardigli, 1988.

The quotes from Reinhart Koselleck that open Section 1.3 are both taken from Koselleck, 2004, p. 87. Regarding the first family of parallel expressions (single technology revolutions), the reflections are drawn from several contributions, but I limit myself to pointing out Etzo and Collender, 2010 on the rhetorical dimension of the penetration of mobile technology in Africa; Margolis and Resnick, 2000 on the fact that some expressions, such as cyberspace revolution, age or normalize and they lose their disruptive appeal. The third family has to do with the mixing together of distinct sectors. Two of the mentioned studies are also valuable historical sources for understanding how the narratives around the digital revolution evolve: Brzezinski, 1970; Nora and Minc, 1978, quotation p. 118. To these can be added Forester, 1981, in which there is also a controversial exchange between Joseph Weizenbaum and Daniel Bell regarding how "revolutionary" the computer really was. The fifth family examines the information and communication revolutions. The *Digital Revolution* entry in Baidu Baike claims: "The digital revolution [is] also known as the Third Industrial Revolution or the third technological revolution." The text mentions Beniger, 1986; Brock, 2003; Headrick, 2000. To remark on how the expression "information and communication revolution" has been used for different technologies is Williams, 1982. Finally, a very funny text that combines diametrically opposed points of view on the information revolution and its effects is Winters, 1998. In the sixth family, the focus is on the concept of digital disruption, transition, and digital transformation. In addition to the Wikipedia entry mentioned, to try to define the concept of digital transformation, see "Digital Transformation", 2016. A definition with a more economic focus is also found in Schallmo et al., 2017.

The quotation from the 2021 G20 declaration is taken from G20, 2021. On the increasingly Asian, and in particular Chinese, digital transformation, see the launch of the ambitious industrial digital transformation project by the Beijing government in 2020 (Savic, 2020).

Finally, on the need to study the historical evolution of the concepts related to digitization, and therefore also of that of the digital revolution, I refer to Balbi et al., 2021.

In Section 1.4 the adjectives "uncertain" and "steerable" are used respectively by Mathews, 2000; and Vitalis, 2016. On the concept of new and always the same, see Morin, 1962. The expression "ever new and always the same" is also used by Merleau-Ponty, 1968, p. 267. On the etymologies of revolution, please refer to Arendt, 1990, and to the aforementioned Koselleck, 2004, in particular the chapter "Historical criteria of the modern concept of revolution."

2

Comparing the Revolution
Past Inheritance, Present Construction

The digital revolution is often described as a phenomenon that started and marked an era, which is then defined as the digital era or age, and its historical significance is often underlined with two rhetorical devices: claiming that the digital revolution is the heir to the great revolutions of the past and that it has fomented or, at least, that it is combined with the revolutions of the present.

On the one hand, it has been associated with and compared to the great revolutions that have marked and changed human history. These include political revolutions, such as the French Revolution at the end of the eighteenth century and the Russian Revolution at the beginning of the twentieth, the scientific revolutions of the mid-sixteenth century as well as the technological revolutions, such as the Industrial Revolution—starting from the eighteenth century—and communication revolutions, such as that initiated by the invention of the printing press in the fifteenth century.

The rhetorical paraphernalia of the digital revolution is enormously enriched by these parallels. In 1984, Apple presented an advertisement in French for its newly launched Macintosh with the telling claim "*Il était temps qu'un capitaliste fasse une révolution* / It was time for a capitalist to make a revolution," where the computer is placed on the same shelf as books written by Mao, Engels, Lenin, Marx, and Trotsky. The digital revolution has its "Robespierre moments" in which open and free markets can have the same propulsive function that the Bolsheviks had in the Russian Revolution (as was claimed during the G7 in 1995) or in which the Fordism that drove the Industrial Revolution has transformed, according to some, into the "Gatesism" of the digital age. The latter underlines how some revolutionary heroes can generate meaningful neologisms. But, even more profoundly, thanks to these narratives, the digital revolution has accumulated a long-term historical legacy. By placing it on the shoulders of giants and identifying it as the heir to great revolutionary traditions that have changed human history in the past, the digital revolution gains immediate recognition and historical importance and is seen as a direct and "inevitable" descendant of those that preceded it. It cannot fail because previous revolutions did not fail (at least on paper); it is irresistible and inevitable just like its predecessors. It even acquires the status of revolution from this glorious past; its future success is predetermined because it is certified by the past. It is the latest in the chronological order of the great revolutions and also the culmination of them, translating many of the earlier revolutionary axioms into a language suitable for the modern world.

On the other hand, the digital revolution is not only described as the heir of the revolutionary past but is also associated with numerous epochal changes in contemporary society, some of which are considered real revolutions. It is often claimed that the digital revolution has functioned as a spark, a necessary precondition to the advent of industry 4.0, the various Arab Springs, the technological or green revolution, the energy transition, among many possible examples. In short, it represents the *ultimate cause* that has initiated and is initiating other revolutions.

The latest iteration of the great revolutions of the past or the ultimate cause of the current ones: defining the digital revolution as a revolution among revolutions is perhaps one of the most powerful narratives surrounding digitalization and the subject we will be investigating in this chapter.

2.1 The Industrial Revolution of Our Time (and Maybe Something More)

Of all the great revolutions of the past, the one most frequently cited and compared to the digital revolution is the Industrial Revolution. Indeed, the digital revolution has often been described as "the Industrial Revolution of our time," both because it is heir to an ideology of unstoppable progress and technological innovation that started in the eighteenth century with the Enlightenment and because it seems to replace the Industrial Revolution, or at least constitute a new stage of it.

That the leap caused by new communication technologies was epochal and comparable to the Industrial Revolution was an idea already supported by Norbert Wiener both in *Cybernetics*, published in 1948, and in *The Human Use of Human Beings*, published in 1950. In the latter text, Wiener devoted an entire chapter to the First and Second Industrial Revolutions, theorizing about how the nascent industry that produced computers would start a new industrial revolution, with political, economic, and socio-cultural consequences as significant or more significant than those triggered by the Industrial Revolution. Wiener claimed that "the possible fields into which the new industrial revolution is likely to penetrate are very extensive, and include all labor performing judgments of a low level, in much the same way as the displaced labor of the earlier industrial revolution included every aspect of human power."

The fact that the digital revolution represents both continuity with and a surpassing of the Industrial Revolution is found in another much-cited "classic" of digitization, Alvin Toffler's *The Third Wave* (1980), the title of which refers to a new historical period generated by the digital revolution and following the Agricultural and Industrial Revolutions.

Other authors have talked about post-industrial society, the information age or era, digital society, and many other similar terms. Certainly, the story narrated in terms of waves or eras is questionable because it hides a teleological and linear vision: in short, it implies that the digital age is the inevitable outcome of all past history or that previous revolutions were a sort of preparation and a prodrome for the digital and, in full revolutionary spirit, that all forms of communication, society, economy, or politics as we have known them will be swept away by the shocking force of the digital revolution. However, whether this new phase is actual or mythological, we have almost universally accepted and internalized the fact that we are living or have experienced a paradigm shift relative to the Industrial Revolution.

Figure 5 The digital revolution among the great revolutions in history.
© CartoonStock.com, cartoonist Fran, uploaded June 5, 2013.

In public discourse, various similarities between the Industrial Revolution and the digital revolution have proliferated and three of these are more often repeated and more relevant than others. In the first place, as already mentioned, both revolutions are said to have generated new forms of society: from industrial or Fordist society, to post-industrial, post-Fordist, information, or simply digital society. The two revolutions have then created new jobs, new relationships between social classes, and new lifestyles. In 2016 Sandeep Aggarwal, CEO and founder of Droom (one of the largest used car trading portals in India), argued that the digital revolution would represent an opportunity for that country very similar to what the Industrial Revolution did for eighteenth- and nineteenth-century Europe: an opportunity that could *radically* change Indian society. In short, the two revolutions are considered "total," that is, capable of influencing the societies of their respective eras as a whole and of initiating new paradigms.

A second similarity between the industrial and digital revolutions concerns the central and symbolic role technologies have played in both of them. As Wiener already pointed out, during the first Industrial Revolution, various technologies had merged and intersected— including the steam engine, vehicles (most importantly ships and railway transport), the manufacture of precision watches, and looms for the textile industry—and that the same would happen in the second. It is precisely the interdependence of this constellation of innovations that constitutes a recurring feature of technological revolutions: it was not so much the steam engine, electricity, or the computer itself that made the respective revolutions but the fact that these technologies merged with others and led to the creation of additional, radical innovations. The same technologies then facilitated various parallels and analogies between the two historical periods, as argued, for example, by John Naisbitt in 1984: "Computer technology is to the information age what mechanization was to the Industrial Age." In 1993 Victor Keegan pointed out that in England, "a national fibre grid . . . could have done for the information revolution what the railways did for 19th century transport." In 1997, two magazines in which the digital revolution has long been presented prophesied that it would produce "a change even more far-reaching than the harnessing of electrical power a century ago" (*The Economist*) and that the computer would become "a transcendent technology—like railroads in the 19th century and automobiles in the 20th" (*Businessweek*). Furthermore, one of the creators of the World Wide Web in 2000 remarked that "[t]he arrival of the web in 1990 was to the Internet like the arrival of the internal combustion engine to the country lane. Internet transport would never be the same again." It is curious that a non-material revolution par excellence, as the digital one has been described, has used and continues to use mechanical, physical, tangible technologies in order to talk about its revolutionary nature.

A third parallel between the two revolutions is the fact that both present opportunities and risks, moreover in some cases very similar ones despite the intervening centuries. As mentioned, the Industrial and digital revolutions are often purported to have led to fundamental changes in politics, the economy, and society. Similarly, the two revolutions have generated familiar concerns about the loss of jobs. The automation of mechanical looms, which also gave rise to the phenomenon known as Luddism, or a form of open dissent that involved the destruction of machines, was seen as threatening jobs at the time in much the same way as artificial intelligence is now seen as a threat to workers in the near future. Similar forms of exploitation also appeared. If in the Industrial Revolution Karl Marx denounced the fact that capitalists/entrepreneurs imposed shifts and economic conditions on workers/wage earners that led to the alienation of the working class, critical digital thinkers like Evgeny Morozov, Christian Fuchs, Jack Qiu, Kate Crawford, and many others have highlighted the persistence of forms of slavery and digital alienation exemplified by home delivery "riders" for companies selling food and other goods, Uber drivers, or Amazon's warehouse workers.

So are the Industrial Revolution and the digital revolution the same thing? Has nothing changed over the last few centuries? Obviously, there are substantial differences between the two historical periods, especially in the scale of the social transformations brought about by the two revolutions. For example, the Industrial Revolution was a more Western phenomenon, while the digital one has a global reach (at least in terms of the rhetoric employed). Furthermore, the digital revolution is presented as having acted and continuing to act faster than the industrial one. It was also discussed which of the two revolutions was the most relevant and, in recent decades, a number of observers have argued that the digital revolution has influenced or will influence human beings in an even more profound and broader way than the industrial one. In a 2018 article titled "Welcome to the Digital Revolution" (titles like this are beyond number), journalist Keven Drum prophesied with certainty that "the digital revolution is going to be the biggest geopolitical revolution in human history." Furthermore, the effects of the digital revolution seemed more immediate than those of the industrial one. For example, in 1997 Raymond Lane, the president of Oracle (entrepreneurs are the true oracles of the digital revolution, and not just because they work at Oracle!), argued that it had taken several decades for the Industrial Revolution to have tangible effects on people's lives, while the digital revolution had been faster. The digital revolution's speed of penetration has resulted in an immediate and almost instinctive recognizability according to David Ferguson, the CEO of Nucleus Financial, who stated in 2012:

> I often wonder if the term "industrial revolution" existed in the consciousness of our ancestors who lived through it. I suspect that the average man and woman who lived in 18th and 19th century Britain were simply doing what they had to do to get by and working much longer and harder in the process than we do today. Now, we have many privileges in relation to our forebears; better healthcare, longer life expectancy and more access to information. This gives us the opportunity to step back and appreciate that we are indeed living through a digital revolution.

According to Ferguson, the awareness of being immersed in a digital revolution, an awareness almost simultaneous to the revolution

"JUST AS WE FINALLY GET THE INDUSTRIAL REVOLUTION DOWN PAT, WE FIND OURSELVES IN THE MIDDLE OF THE ELECTRONIC REVOLUTION."

Figure 6 You need time to metabolize revolutions!

© CartoonStock.com, cartoonist Sidney Harris, uploaded January 17, 2012.

itself, is therefore one of the characteristics that distinguishes this revolution from the others that preceded it, *primarily* the industrial one. The same awareness was also set to music by the rap group Knights of 88 (the revolution also has its bards), in a 2021 song entitled *Digital Revolution*, in which they ask ironically "Anybody heard of the digital revolution"?

In short, if the other revolutions were identified and understood only decades after their actual appearance (the case of the "scientific revolution" of the sixteenth century, which was only labeled as such starting in the 1930s), the digital one was or is immediately recognized as a revolution, naturally also in order to market and promote it. "We are all very privileged to be in this great information revolution, in which the computer is going to affect us very profoundly, probably more so than the Industrial Revolution," Michael Dertouzos, director of MIT's Laboratory for Computer Science in Boston, told NBC in 1983; privileged and proud to be part of an epochal change, to experience a paradigm shift in a conscious way. We certainly cannot complain.

2.2 Communicating Better, Again: From Gutenberg to Zuckerberg

The digital revolution has also been described as the heir to and an improvement on various previous communication revolutions. In recent decades, dozens of scholars, journalists, politicians, entrepreneurs, and opinion makers have argued that new communication technologies revolutionize human habits and in many of these "revolutionary histories" of communication, there is a recurring pattern: new media and new forms of mediated interaction substantially change the way we communicate and the social order, generating a new communicative era distinct from the previous ones until the appearance of another disruptive medium initiates a new paradigm (to use a term evidently inspired by Thomas Kuhn's work on scientific revolutions) and a new epoch. There are of course also authors who have criticized revolutionary vocabulary and periodicity, arguing that "revolution" is a "cliché" and an "overused term," that there are forms of "conservative revolutions" that preserve the previous status quo, or that when discussing communication it's better to talk about evolution or coevolution.

But the revolutionary discourse is more fascinating because it provides simple stories both to explain technological substitution and to propose periodizations for "great eras" that are easy to remember. Among the various communication technologies that have become part of the rhetoric of the disruptive revolution are the alphabet, writing, printing, letters, electric media (such as the telegraph and telephone), visual media (for example photography and cinema), radio and television, and, of course, electronic and then digital media. Each of these communication tools has been seen as a bearer of political, economic, and social consequences because often an information revolution is capable of subverting not only the extant communication methods but also the lifestyle and interrelations of entire societies and, potentially, of humanity itself. Among the most significant changes, repeated for decades and perfectly adaptable to the digital revolution, the most significant is perhaps the fact that the new communication technologies (be it letters, the telegraph, or the Internet) almost always allow human communication to be sped up, eroding time and space, bringing people closer together (a famous claim reappropriated by Facebook in the late 2010s), connecting people (to quote Nokia in the 1990s), allowing them to talk to each other, exchange and receive information more and more rapidly up to the point where that communication is instantaneous. The inevitable acceleration of new media involves a series of transformations which are also often repeated. More communication should, for example, correspond to more understanding, so much so that new media are often seen as a way to create peace among people. More communication possibilities should lead in a "natural" way to the development of freer, more democratic, and less hierarchical societies. In this respect, the Internet is just the latest example of a form of communication seen as horizontal and with democratizing potential. Or, again, since the various revolutionary forms of communication (letters, telephony, radio and television, social media up to and including the Metaverse) allow you to travel virtually faster and faster, this should reduce the propensity to use other means to travel physically. Finally, there should be an inevitable trend towards more effective, economical, easy-to-use forms of communication, which naturally culminates with digital technologies but which has its own historicity: the telephone is better than the telegraph which, in turn, is better than the mail and so on; or the smartphone is better and smarter than 1990s mobile phones which were a stunning breakthrough compared to landline telephone. In other words, the "social effects" of communication revolutions are (or should be) legion, but what is most surprising is their recurrence, which is to say the fact that they are repeated centuries later and for every communication "revolution" and that, each time, they are presented as completely new, unexplored arguments which perfectly explain the spirit of the latest and most powerful technological breakthrough.

The digital revolution has been described as the last in a long line of communication revolutions and, for this reason, it has also inherited most of the stereotypes of the previously mentioned revolutions. Surprisingly, the "printing revolution" is the communication revolution most often compared to the digital revolution. Just as was the case with the comparison between the Industrial Revolution and the digital revolution, the latter was also seen as a new phase, or even the conclusion, of the printing revolution. Over the decades there has been an incessant flow of forecasts regarding the disappearance of books, newspapers, and magazines, or, by extension, of paper. Others have predicted the breakdown of monopolies held by large publishing houses together with the emergence of self-publishing and printing, and the end of retail bookstores. And, once again, these come with the rhetorical parallels: the transition from Gutenberg' to Zuckerberg's era is not only one of the most repeated slogans in multiple sources and evidently favored by the assonance between the two surnames.

According to historian Jérôme Bourdon, in recent decades the parallel between the two revolutions have become something of a cliché even for media historians. In fact, even before the scholars, the idea that the two revolutions had strong similarities was purveyed by some digital gurus. In his well-known book, *The Road Ahead*, published in 1995, Bill Gates wrote: "The information highway will transform our culture as dramatically as Gutenberg's press did the Middle Ages." In *Internet Galaxy* (2001)—which evidently referred to Marshall McLuhan's *The Gutenberg Galaxy*—sociologist Manuel Castells argued that the Internet would have an impact on contemporary societies similar to that of the printed book on fifteenth-century European societies. The analogy has also been adopted by politicians. In 1995, Singapore's Minister of Information, George Yeo, inaugurating an online services company in the country, declared that the digital revolution "will have an impact on human society as profound as the introduction of printing in Europe in the 15th century." And in 2000, a Special Commission of the Swiss

"It's exhilarating to know we're riding the crest of the digital media revolution."

Figure 7 At the crest of the digital revolution, but keeping old habits.
© CartoonStock.com, cartoonist Bill Borders, uploaded November 10, 2019.

National Council wrote that "this revolution will bring about changes as fundamental as those that occurred when Gutenberg invented the movable type printing press."

But for what reasons are the two revolutions compared and comparable? A fairly thorough answer can be found in a 1998 paper from a famous US think tank, the RAND Corporation. In fact, in both revolutions three comparable conditions have arisen: new technologies such as the printed book and the Internet have allowed for the expansion and intensification of communication between human beings; unprecedented ways of storing, distributing, updating, and searching for information; and finally, a series of unpredictable consequences generated by new technologies such as the Protestant Reformation or the scientific revolution for the printed book. In short, both the printing press and digitization are two transformative revolutions for the whole of society, with long-term consequences and effects on the way information is communicated and manipulated, and in fields far removed from communication. Systemic revolutions, which in turn instigate other revolutions, will be discussed in Section 2.3 of this chapter.

In recent decades, a debate has also been sparked regarding the revolutionary nature of the printing press revolution itself, a debate that provides food for thought when it comes to reconsidering the "revolutionary nature" of the digital revolution. The most bitter and interesting academic dispute on the subject took place in the pages of *The American Historical Review* in 2002 when two of the greatest scholars of the printing press, Elizabeth Eisenstein and Adrian Johns, expressed opposing views. The first, author of key texts such as *The Printing Revolution in Early Modern Europe*, published in 1983, has always maintained that printing was a real (and unintended) revolution for fifteenth-century society and caused a sharp break between a pre- and post-print society. Johns, especially in *The Nature of the Book*, published in 1998, supported the opposite thesis: that the revolutionary nature of the printed book was a discursive construct, born alongside the French Revolution, which was useful to evoke a disruptive past by which one could be inspired but which was imposed only during the nineteenth century. Johns, who goes so far as to call the printing revolution an "ideology," as I do with the digital one, argues that the printing press did not revolution-ize the society in which it emerged but that its revolutionary nature was retrospectively imposed only during the French Revolution to serve a rhetorical function and to support partisan interests. It is a disruptive thesis but one whose echoes, especially regarding the ideological use of the digital revolution, can also be found in the pages of this book.

2.3 A Revolution Intercepting and Founding Contemporary Revolutions

The digital revolution is not only the heir to and the culmination of the great revolutions of the past, but it has also often described as the spark, the cause, the trigger, or even the *sine qua non* that has made or is mak-ing some of the most significant contemporary revolutions possible. To paraphrase a well-known expression of Karl Marx, it could be said that the digital revolution is recent history's locomotive.

In Chapter 1 it was mentioned that the phrases digital revolution, and increasingly digital transformation, are considered synonyms for the Fourth Industrial Revolution and therefore that the digital one not only supersedes but is also a new phase of the Industrial Revolution. However, there's something more: industry 4.0 would not be possible without the massive use of the products of the digital revolution (from computers, to high-speed connections, to robots, to AI) and therefore the Fourth Industrial Revolution is not only synonymous with the digital revolution but also one of its products.

The philosopher Luciano Floridi also talks about the *fourth revolution*, but he is referring to a paradigmatic leap in the perception of the self and not a change in industrial production models. According to Floridi, human beings have experienced four major revolutions and each of these has generated uncertainties and changes in previous-ly established perspectives. In his view, the first revolution was the

Copernican one, in which humans and the planet they inhabited and inhabit ceased to be at the center of the universe and instead assumed a peripheral position relative to the Sun. The second revolution was introduced by Darwin's evolutionary theory, thanks to which human beings have ceased to consider themselves the great exception in the animal kingdom, naturally the most intelligent and advanced, and have instead found themselves to be a simple evolution of primates, animals among animals. Floridi associates the third revolution with Freud's psychoanalysis. If many of the actions we perform are the result of the unconscious, we can no longer consider ourselves rational Cartesian subjects and completely aware of every aspect of our personalities. Therefore, even the deepest self has become opaque and difficult for humans to know. Finally, with the fourth revolution, which Floridi traces back to Alan Turing (and therefore to one of the patriarchs of the digital revolution, as we will see in Chapter 4), it is human *nature* itself that is jeopardized. A test devised by the British mathematician in 1950 to determine whether a machine is capable of exhibiting "intelligent" behavior, known as the Turing Test, taught us that machines and humans can be confused with each other and therefore intelligence is no longer just a human prerogative but can also be artificial and digital. This umpteenth disorientation has initiated a fourth revolution, in which humanity is forced once again to abandon its certainties, to even change its own nature in a move towards an intangible "informational" dimension (therefore one of the cornerstones of the digital revolution re-emerges, the dematerialization from atoms to bits) and in which one of the main criteria of existence is digital interaction and communication. So, from this perspective, the digital revolution is causing a paradigm shift and a true transformation of human nature itself.

Among the many other examples of revolutions caused by and intertwined with the digital one, I would like to mention three more. The first is that of the revolutions of the so-called Arab Spring. We're talking about a vast phenomenon of protests, revolts, and genuine

political-social revolutions that affected various North African countries between 2010 and 2012 and which were resolved in different ways,
from the overthrow of some authoritarian governments in favor of
other, more democratic or even more despotic ones, to repression, to
political chaos. The revolutions of the Arab Spring were renamed *Twitter
Revolutions*, or protests in which Twitter played a crucial and instigating
role both because it was the platform through which the revolutionaries communicated with each other and because it represented a public
space in which to inform the rest of the world about the reasons behind
the protests. Although the centrality of Twitter has been contested by
both media scholars and journalists (it is enough to remember that
in 2010, only 1% of the population of the countries in question had a
Twitter profile), the idea that social media favors political revolutions
is still popular today. Indeed, the semantics of Twitter Revolutions has
expanded in recent years, as evidenced by the Wikipedia page dedicated
to them which includes, for example, the protests in Moldova in 2009,
those in Iran between 2009 and 2010 (also known as *Facebook revolutions*)
and in Ukraine in 2013, as well as "the storming of the United States
Capitol" in early 2021, spurred by the tweets of the then-incumbent
President, Donald Trump. In other words, one of the most common
and, at the same time, most misleading narratives wants to portray the
digital revolution as a trigger and a tool so indispensable to popular
uprisings that they been renamed Twitter and Facebook revolutions,
revolts directly provoked by digital platforms.

The digital revolution is also intertwined with many of the contemporary technological revolutions and there is a person who, perhaps
more than anyone else, has been the glue between the different revolutions. The South African entrepreneur Elon Musk, in his various
and sometimes crazy exploits, has brought together several revolutions and I am not talking about the acquisition of Twitter itself: Tesla,
Hyperloop, and SpaceX all refer to a revolutionary vision of mobility
(even if in reality it is often old ideas re-presented as innovative) such

as autonomous driving, electric power supply, or the colonization of space. SolarCity aims to bring about a revolution in the production and consumption of electricity. Starlink, a constellation of satellites to provide Internet connection to the entire planet, promises a revolution in Internet access, seen as a universal good. It goes without saying that all these projects are closely connected to digitization, the presence of broadband networks capable of transporting data, and, in essence, they develop the digital revolution or transformation in new ways, making it the hub around which other changes, considered epochal by contemporary society, revolve.

A final example of an ongoing revolution in which digitization plays a fundamental role is the so-called *green revolution*. Between the 1940s and the 1970s, this expression meant the use of new technologies for agricultural production: from the use of genetically modified seeds and organisms to fertilizers, to new means of transport that allowed a significant increase in agricultural production in different regions of the world. Today we tend to extend the concept of green revolution to include the so-called *green economy*. Also due to the challenges posed by climate change (probably one of the few human transformations that can compete with the digital revolution in terms of rhetorical power on a global level), various governments, international bodies, and private companies are promoting an ecological transformation or transition towards a model of human society that reduces the use of carbon-based fuels in favor of renewable and sustainable energy resources. Digital transformation is often combined "naturally" with the green energy transition because it seems to be the tool through which polluting emissions can be reduced, for example by limiting physical movement, optimizing industrial and private consumption, as well as monitoring logistics and transport systems. Also in this case, the fact that several studies have highlighted the degree to which the digital world pollutes (from energy consumption for servers and supercomputers or that needed to produce cryptocurrencies, to the rapid obsolescence of devices such as personal computers or smartphones which contributes to increasing e-waste and digital waste) has, so far, not affected this narrative. Perhaps the most powerful example of this is the European Union's *Declaration on A Green and Digital Transformation*, signed by 27 member states in March 2021, whose main objective is to make Europe the world leader "in the green digital transformation" (with this phrase the two transformations—digital and green—are definitively merged). The Declaration's objectives include the intent to:

> [i]ncrease the energy efficiency by modernising connectivity and electricity distribution networks, enabled by the digital transformation of systems and processes . . . Speed up the green and digital transformation of public services by making public services available on line in an inclusive manner (education, healthcare, agriculture and e-government services) and actively facilitating telework during the pandemic and beyond, including by accelerating the roll-out of energy efficient ultra-fast broadband networks.

From these words it is clear how the politicians (and not only European ones) have absolutely no doubts about the transformative power of digitization even in confronting what will be the most important challenge of the coming decades. Furthermore, and once again, these are recurring arguments. For example, rereading the previously mentioned 1993 document from the US Clinton administration, *The National Information Infrastructure* plan, one immediately notices that the services and sectors transformed by the digital revolution are always the same—just as the political and economic rhetoric has also been repeated for several decades now—with only minimal variations. Moreover, both the European Declaration of 2021 and the US plan from the early 1990s have had a direct economic impact. In fact, in the European *Next Generation EU* investment plans, most of the funds the EU allocated to the Member States to reinvigorate economic activities after the COVID-19 pandemic *must* be spent on the digital and energy transition. In short, the digital revolution also imposes business models on the public sector.

As we are often reminded, the digital revolution is at the center of all the most important contemporary revolutions and, therefore, in order to understand the other revolutions and even our own times, we must first understand it. After all, the great revolutions of an era must be analyzed together and only by looking at them in their mutual interactions can one truly understand a historical period. From this perspective, the adjective *ultimate* was used at the beginning of this chapter: the digital revolution is often considered the ultimate cause and the spark necessary to instigate other revolutions and more generally social change. Indeed, in some ways the changes in contemporary society can only be caused by, or at least are strictly related to, the digital revolution.

Endnotes

On the comparison between the digital revolution and other revolutions, André Vitalis argues that the digital revolution is described as the heir to past revolutions such as the steam engine or electricity and, also thanks to this frame, has acquired a "sense of historical inevitability" that puts it "at the center of the discussion about the future" (Vitalis, 2016, p. 90). Also at the beginning of this chapter, I give three examples of rhetorical tools that the comparison with past revolutions provides to the digital one: for the French Revolution, the Robespierre moment is quoted in Jenkins, 2012; for the Russian Revolution, the comparison between free markets and Bolshevism is in MacKinnon, 1995; for the Industrial Revolution, which will then be the subject of an in-depth discussion, the transition from Fordism to Gatesism is theorized by Tremblay, 1995.

The literature that compares the Industrial Revolution to the digital revolution, on which I rely for Section 2.1, is substantial. Among the most interesting contributions, see Freeman and Louçã, 2001; Kleine and Unwin, 2009; Kulwiec, 1994; Mathews, 2000. The claim that the digital revolution is "the Industrial Revolution of our time" is supported, for example, by Kaufman, 2012, and by the already quoted Vitalis, 2016, p. 88. In this section I also quote Toffler, 1980, and Wiener, 1948, 1989 (there is also a quotation from this latter book at p. 159). On the digital revolution as an Indian Industrial Revolution, see Borpuzari and Bhattacharya, 2016. On the technological dimension of the two revolutions, see Perez, 2004. Carlota Perez has also been concerned for a long time with the technological dimension of the digital revolution, especially in the 1980s. The technological parallels between the two revolutions are mentioned in Gillies and Cailliau, 2000, p. 1; Keegan, 1993; Madrick, 2014 (which takes up the 1997 articles published in *The Economist* and *Businessweek*), and Naisbitt, 1984. On the digital revolution as an even deeper change than the industrial one, see Oracle's president Lane, 1997; "Are We Aware We Are Living," 2012, p. 2425. Michael Dertouzos is cited in Winner, 1984, p. 582. The song Digital Revolution by Knights of 88 is on their 2021 album that reflects on digitization, titled *Tech Rap: Jumped Off the Block Onto Blockchain*. Finally, historian Porter, 1986, argues that the scientific revolution was recognized as such only hundreds of years after it took place.

At the beginning of Section 2.2, I argue that different means of communication have been narrated as "revolutions" and various handbooks and works dedicated to the history of communications adopt this perspective composed of grand periodizations and disruptive changes. Interesting as a paradigmatic reference for many histories of the revolution is Kuhn, 1970. A classic on the study of communication revolutions is Behringer, 2006, which should be read together with John, 1994. De Sola Pool, 1982, calls the term revolution a *cliché* and refers to it as an overused word. Fickers, 2012 argued that television is an example of a "conservative revolution." There are many texts that speak in favor of an evolutionary approach, like Winston, 1998 and Scolari, 2023, or considering the role of continuities even more relevant than radical changes in media studies, see for example Driessens, 2023.

There is also a great deal of literature on the repetition of the alleged effects of the information and communication revolutions. On the reduction of distances, see Cairncross, 1997. On speeding up the relationship between transport and communications, see Carey, 1989. On the growing efficiency and presumed linearity in the development of digital forms of communication, see the popular (at least in media studies) concept of "deep mediatization" in Hepp, 2020, and for a critique refer also to Bourdon and Balbi, 2021.

On the digital revolution as surpassing the printing revolution, the literature is once again extensive. I would like to highlight Epstein, 2008 for a decidedly apocalyptic vision. The expression "from Gutenberg to Zuckerberg" is quite overused in conferences, journalism courses, newspaper articles, blogs, and, of course, books. The best-known book that uses this title, even if it's just an excuse to talk about the Internet's "disruptive innovation," is Naughton, 2014. The comparison between the printing revolution and the digital revolution is always a *cliché*, according to Bourdon, 2018. Among the proponents of this parallelism, the following are cited: Gates, 1995, p. 9; and Castells, 2001. The quotation from Singapore's minister is in "New On-Line Service Unveiled by Singapore," 1995, p. 14. Also cited is the *Report of the Special Commission of the Swiss National Council, Legislature Program 1999–2003*, 2000, p. 5187. Regarding the main reasons why the printing revolution and the digital revolution are often compared, see Dewar, 1998. For the debate on the revolutionary nature of the printing press, see Eisenstein, 2002 and Johns, 2002. Two other works of these scholars are quoted: Eisenstein, 1993 and Johns, 1998, which, on p. 374, reads: "The ideology . . . of a printing revolution was . . . not only a result of the French Revolution: it was perceived as . . . necessary in order to render that revolution both permanent and universal."

I open Section 2.4 by paraphrasing Karl Marx when he argues that "Revolutions are the locomotives of history" (1976, p. 120). On the industry 4.0 and its ideological dimension, see for the German case Fuchs, 2018. I then quote Floridi, 2014, Chapter 4. Various sources disprove Twitter's role in the Arab Spring revolutions, notably see, on the academic side, Curran, 2012; on the journalistic side, Beaumont, 2011. The play on words between a political revolt and a revolution brings to mind the dialog between Louis XVI and the Duke of La Rochefoucauld-Liancourt on the night of July 14, 1789: the King exclaimed "*C'est une révolte!*" and La Rochefoucauld-Liancourt corrected him: "*Non, Sire, c'est une révolution*" (quoted in Arendt, 1990, p. 41). On climate change and digitalization, you can find the Wikipedia page with the definition of the *Twitter Revolution* and online the European Green Digital Coalition, 2020. More recently, the digital revolution at large has been accused of polluting as well as wasting large amounts of energy and rare-earth elements for producing digital devices and maintain data centers and servers: see, for example, Crawford, 2021; and Pitron, 2021. On Elon Musk and digital transformation there are dozens of enthusiastic articles and books and as many critics, especially regarding the business models adopted by the entrepreneur, particularly after the acquisition of Twitter in 2022. But perhaps the most interesting source is the almost hagiographic documentary *Elon Musk: The Real Life Iron Man*, 2018 (presented with the following words: "Discover the meteoric rise of Elon Musk, the man who is transforming the way we think about travel technology through electric cars, the Hyperloop, and revolutionary ideas on how we live through artificial intelligence and colonizing Mars"). In 2021, Netflix also co-produced *Countdown: Inspiration4 Mission to Space*, on the SpaceX mission, to validate a media interest in Musk and his revolutionary exploits.

3

Thinking About the Revolution
The Mantras

What are the main characteristics that make the digital revolution revolutionary? What revolutionary reflections and thoughts is it heir to, even unknowingly? To answer these questions, which are at the core of a book that aims to analyze how the digital revolution has been narrated, this chapter will trace five of the most significant and repeated mantras with which this revolution has been described (disruptive, total, irresistible, linked to the future, and permanent). These themes have repeatedly emerged from the sources, even in documents produced at different times and by different cultures, and, surprisingly, can also be considered as voluntary or involuntary legacies of and in dialog with some of the classics of revolutionary thought. After trying to define it in the first chapter and comparing it with other revolutions in the second, tracing the legacies of "classical" revolutionary thought in the frequently repeated formulas and slogans of the digital revolution, which have become almost automatic, is a further step towards understanding its ideological scope.

3.1 A Disruptive Revolution

The mother of all revolutionary arguments is the one which claims that the digital revolution (as well as the others that preceded it) constitutes a break from the past. As already mentioned when citing the

Wikipedia entry in Chapter 1, the main disruption in the history of the digital revolution seems to be the transition between analog and digital, between atoms and bits. The digital revolution is "disruptive," a term that comes from Latin *dirumpere* which means to break, split, or interrupt, and is often used in the context of war to indicate a violent deflagration. According to revolution theorist Karl Griewank, this is one of the three typical characteristics of political revolutions, namely, "a process which is both violent and in the nature of a sudden shock—a breaking through or overturning, especially as regards changes in the institutions or state and law."

And is the digital revolution violent? Apparently not, also given the notable absence of counter-revolutionaries who are willing to die fighting against it. But in reality, there are at least two ways in which the digital revolution can be associated with violence and even death. The first relates to the disappearance of old media and previous forms of communication. With the advent of the digital revolution the following that in reality are all still extant, were and are supposed to have been eradicated: in no particular order books, newspapers (and paper in general), the television, analog photography, the music industry, cinema, and many other forms of communication. In short, the digital revolution is a wannabe killer of the analog past. However, there is a second reason, which the most popular narratives tend to hide and that is decidedly more dramatic and real: in the name of the digital revolution people die in order to extract, often with their bare hands, rare materials in Third World countries; people die because they can't cope with the hectic pace of producing digital devices (in China, for example, some worker suicides have been reported in factories where Apple's iPhones are produced); people die because the digital objects we so frequently replace are piled up in huge landfills, particularly in some African and Asian countries, and, due to the toxic materials they contain, create long-term pollution issues. In this sense, the digital revolution is a break that is not only violent but even lethal.

The stories that highlight the revolutionary and disruptive nature of the digital revolution take on different forms and, as already mentioned, the new eras or phases that have started thanks to digital are beyond number. An interesting example is related to the economy. Since at least the 1970s, if not earlier, we should theoretically have been immersed in a "new economy" with business models that are totally different compared to those of the past. And of course, these come with their related acronyms, such as NEBM, which stands for New Economy Business Model, "new economic paradigms" (a concept that Alan Greenspan, then chairman of the US Federal Reserve, championed in the 1990s), and "new forms of capitalism" (here we have literally gone wild in coining slogans such as information capitalism, cognitive capitalism, neo-capitalism, digital capitalism, surveillance capitalism, and many others). In short, as already mentioned, the digital revolution continuously invents labels that certify its revolutionary and transformative status, even when it is not clear what to call a new phenomenon, nor is it certain that it will be all that disruptive. In the economic field, one of the most symbolic definitions is that of "friction-free" capitalism. Theorized by Bill Gates himself, this new phase of capitalism fulfills or would fully fulfill the dream of Adam Smith (it's no coincidence that he is the chosen cult-author of liberalism as he was among the first to support the self-regulatory capacity of the market through the metaphorical "invisible hand"), allowing vendors and consumers to have all the necessary information on the market and thus be able to make fully informed choices. In particular, it is (or would be) the Internet that creates this perfect world in which product prices can be compared globally and everyone can choose, freely and with all the information at their fingertips, what to buy. In short, the digital revolution also changes the law of supply and demand.

Once again, the disruption of the digital revolution is explained by drawing various parallels with radical and disruptive inventions of the past. The invention of fire, for example, has been one of the most popular. In 1992, the French government's technological consultant Jacques Attali wrote that the impact of the digital revolution "will be more akin to the discovery of fire by [our] early ancestors, since it will prepare the way for a revolutionary leap into a new age that will profoundly transform human culture." In 1996 the activist John Perry Barlow, a real militant of the digital revolution, authored the well-known *A Declaration of the Independence of Cyberspace*. A year earlier he had written: "With the development of the Internet, and with the increasing pervasiveness of communication between networked computers, we are in the middle of the most transforming technological event since the capture of fire. I used to think that it was just the biggest thing since Gutenberg, but now I think you have to go back farther." However, the most well-known parallel with the invention of fire can be found in the inaugural issue of *Wired*, in early 1993, in which editor-in-chief Louis Rossetto wrote a manifesto indicating the reasons that led to the creation of the magazine:

> Why WIRED? Because the Digital Revolution is whipping through our lives like a Bengali typhoon—while the mainstream media is still groping for the snooze button. And because the computer "press" is too busy ... to discuss the meaning or context of social changes so profound their only parallel is probably the discovery of fire.

The famous metaphor of the Bengali typhoon is indicative of a revolution that could have wiped out all the media and the previous universe, leading to radical social changes. *Wired* therefore set out to be an interpreter and to tell the story of the revolution, giving voice to the new digital generation that would inevitably take the place of the analog one linked to sleepy traditional media. And the magazine succeeded in its intention of becoming the bard of the revolution, as will also be shown in Chapter 4.

The quotations that contain parallels with the invention of fire are prototypical of two ways in which the digital revolution is usually viewed: it is a transformative revolution (the term appears in the first two quotations) for human society and it is a spectacular, profound revolution. Fire is also significant because it refers to one of the most important paradigm shifts in the history of humanity: from a society close to the animal kingdom (where food was eaten raw, there was no source of external warmth, and the nights were dark) to a human society in which fire allowed people to cook food, warm up, and see at night. Finally, fire as a metaphor for digitalization in general has enjoyed great success and longevity, and not only for the digital revolution. In 2021 the CEO of Alphabet, and therefore also of Google, Sundar Pichai said that artificial intelligence "will have a more profound impact on humanity than fire, electricity and the Internet" and that it would radically change human lifestyles. In fact, this same mantra is repeatedly taken up, time and time again, when technologies change.

3.2 A Total Revolution

"Total Revolution" (Sampoorna Kranti) was the name of a student political movement led by Jayaprakash Narayan in India in the 1970s. The movement's objective was to oppose corruption in the central government. But the expression "total revolution" can also be applied to the digital revolution, thus constituting another point of contact with the Indian tradition, for at least three reasons: the digital revolution involves all activities and sectors of interest to humans, it is universal, and it is happening in a similar way around the world. As mentioned in the introduction, it is a revolution that has taken on nuances that are all-encompassing because it wants to provide an *official ideology* for understanding and explaining the world and all its phenomena.

The total, in mathematical language, indicates the sum of different numbers. If we use this definition, the digital revolution can be defined as *total* because while changing the various sectors of human activity separately, their sum generates a broader, overall change, an overarching, transformative result. In recent decades, thousands of books, articles, and public talks have unequivocally (another adverb overused in descriptions of the digital revolution) established that agriculture, art, design, the economy, fashion, finance and monetary systems, industry, education, engineering, law, media, religion, politics, physics, health-care, transportation, and dozens of other human sectors have been transformed by the digital revolution. If we were to go back a few years and look at these sectors before the digital revolution, we would no longer recognize them. In general, it has been said that these sectors have accelerated, are more efficient, automated, have long since broken down established hierarchies, and have undergone various other transformations—even if this has occurred at different times and in different ways. Thanks to the digital revolution, human beings will be able to access desired services anywhere and at any time, they will work less, they will have more free time available, they will be more efficient (these are some of the most often repeated promises). In some cases, nostalgic or critical positions have also emerged, which lament the loss of our ability to concentrate, think, and the uniqueness of individual products; however, even those who take these positions don't call into question the (negative) impact that digitalization has had on the sector or the existence of a digital revolution.

The most significant change promised by the digital revolution is the transformation of human nature, a process described as inevitable—and after all, one of the meanings of the adjective "total" refers to something complete and absolute. In 1982, *Time* magazine dubbed the computer the person of the year, the first non-human to receive this title, commenting: "The 'information revolution' that futurists have long predicted has arrived, bringing with it the promise of dramatic changes in the way people live and work, perhaps even in the way they think." In 1998, the oft-cited *Wired* wrote: "We stand on the verge of being able to change the human race." In 2014, writing for the *Wall Street Journal*, Mark Zuckerberg stated that "[t]here have been moments in history where the invention of new technology has completely rewired the way our society lives and works . . . In the coming decades, we will see the greatest revolution yet, as billions of people connect to the

Internet for the first time." In 2017, the World Economic Forum in collaboration with the IT services and the consulting company Accenture launched the so-called *Digital Transformation Initiative* with these words: "The world is being transformed by new technologies, which are . . . changing the way people live and work. Digital transformation . . . has immense potential to change consumer lives, create value for business and unlock broader societal benefits." These are just four examples, among the hundreds possible, of radical statements about the total effects that the digital revolution would have on the *human race itself*, on our way of working, living, and thinking (three identical words repeated decades apart in the articles just quoted). There are also examples on the opposite but complementary side: the digital revolution can be considered one of humanity's worst mistakes because it has the potential to increase divisions and inequalities, even in terms of cognition, and so to change humans in negative ways.

That human nature itself is at stake and that it could undergo a radical transformation is also supported by three intertwined doctrines that, in recent decades, have become increasingly popular: transhumanism, which hopes to eliminate some unwanted aspects of the human experience, such as illness and aging, through digital technologies; posthumanism, which proposes a physical and mental transformation of human beings into new hybrid beings (human and nonhuman) thanks to digital enhancement; and the technological singularity involving the advent of an intelligence superior to human intelligence, generally artificial and digital, which will cause humanity itself to lose its position as more intelligent than what it has created and which refers back to Floridi's fourth revolution discussed in Chapter 2. Regardless of whether transhumanism, posthumanism, and technological singularity actually happen or not—for some they are already present in contemporary society, for others they cannot ever come into being—it is crucial to note how the digital revolution (or transformation) is always described as potentially capable of modifying human *nature*, perhaps the most total and radical transformation possible.

Figure 8 Finally, the real meaning of digital transformation.
© cartoonstock.com, cartoonist Royston Robertson, uploaded September 24, 2019.

The digital revolution is also total because it involves the entire world and is often presented as a global revolution. This is the second meaning of the adjective on which I'd like to focus, and it is a mantra that, surprisingly, the digital revolution seems to borrow from the classics of communist revolutionary thought such as Marx and Engels: the proletarian revolution, and by extension many of the true social and total revolutions, cannot be carried out in a single country but are necessarily international. When we shift from "Workers of the world unite!"

to "Consumers of the world, join the revolution!" there are obviously substantial differences. But perhaps the digital revolution is the most evident and, as mentioned, surprising realization of the idea of an international revolution because it has taken root in various countries around the world, with narratives and arguments that are essentially identical. For example, the RAND Corporation, the US think-tank, regularly issues reports on the state of the global information revolution and evaluates its impact in different areas of the globe as well as in socio-cultural sectors such as technology, finance, and politics. The largest international organization that deals with communication, the International Telecommunication Union (ITU), produces reports on how to make the digital revolution revolutionary for developing countries as well as developed states and therefore tries to guide it, adapt it, and in fact extend it to the entire world. These and other reports from similar institutions often underline how the revolution is taking place, but there is still a way to go (especially to bring some countries to the appropriate level of digital advancement, as if there were some prerequisite standards to be achieved), and the decisive changes will take place in the coming decades. These are, in short, some classic mantras intertwining the present and the future of the digital revolution that will be analyzed later in the chapter.

However, there are two additional arguments that are often repeated regarding totality as a global whole. The first: the digital revolution is explicitly defined as "universal" and as one of the few solutions to the most important problems that have afflicted all humanity for millennia. It was Nicholas Negroponte himself who—in the aforementioned Bible of the digital revolution, *Being Digital*—claimed that "many intellectual movements are distinctly driven by national and ethnic forces, but the digital revolution is not. Its ethos and appeal are *as universal as rock music*." Although fire is certainly a human universal, perhaps rock music is an even better example of a product of human ingenuity that is appreciated everywhere and in different cultures; something as "easy" to understand as the digital revolution is (or should be). In short, to refer again to the conceptual historian Reinhart Koselleck, like other revolutions, the rhetoric of the digital revolution is that of the "collective singular": all human beings buy the products of the revolution and believe that these products are made to suit their needs and passions, but by purchasing and using these tools they become part of a community that sees, in the digital revolution itself, its own reflection.

A second, even more powerful narrative states that all humanity will be able to solve its problems *thanks to* digital technology. This rhetoric is once again theorized by a digitization guru, Google's CEO and cofounder Larry Page, who, in addition to being a strong supporter of the digital revolution (whose noted and oft-quoted statement: "we need revolutionary change, not incremental change," is of uncertain origin, as is often the case), believes that, thanks to it, "we could probably solve a lot of the issues we have as humans." In addition, advertising has gone wild in identifying humanity's most urgent problems which the digital revolution could solve: poverty (digital could break down inequalities and develop new profitable markets); hunger (Gates himself has continually repeated this mantra recent decades); climate change (this rhetoric was already addressed in Chapter 2); and of course pandemics (after all, with the spread of COVID-19, digital companies have facilitated communications and enabled home delivery of products even when our mobility was limited). However, one of this trope's most powerful narratives is that of life-enhancing technology, technology that improves our lives and that even considers the digital revolution thaumaturgical: encouraging physical movement, reminding us to eat healthier, to rest, or to take care of ourselves, optimizing and bringing public administration closer to the citizen-user, creating less pollution by reducing travel and therefore CO_2 emissions. These are all examples of solutions to our daily problems offered by digitalization. Large or small solutions (or "solutionisms," as Evgeny Morozov would say), which involve all citizens, should make it possible to start doing as Google's most popular slogan suggests: "making the world a better place." And this is another classic mantra of political revolutions that the digital revolution makes its own. Just as many of the revolutions of the past have had the goal of achieving a collective good and have often set out to reduce or eliminate poverty, the digital revolution also seems to have the ultimate and very ambitious goal of achieving a greater and collective good: improving lives, solving most of humanity's problems, and making the world a better place. It is an objective that is as total and as complex as it is ideological.

3.3 An Irresistible Revolution (and its Three Laws)

Another revolutionary feature of the digital revolution is its irresistibility, something which as two primary meanings. First, the digital revolution seems irresistible because it is "too attractive and tempting to be resisted," because it "entices, attracts, captivates, and appeals to" contemporary societies. In particular, billions of people are fascinated by the revolution's corporate gurus and CEOs and their—all very similar—"mythological" lives. Even more fascinating are the digital devices that their companies produce: computers, latest generation phones, music players, wearables, and several other products which are all launched with great fanfare during high-tech fairs and then sold in stores located in the most strategic locations in major cities across the globe. Those two subjects will be addressed in Chapter 4 in the sections dealing with the evangelists and relics of the digital revolution.

But a second meaning of "irresistible" is perhaps more interesting: the digital revolution is "too powerful or convincing to be resisted," it is "inexorable," "inescapable," or, what's more, "imperative." There are not many enemies of the digital revolution, something that will be addressed again in the section of Chapter 4 dedicated to heretics and infidels, and very few of them have argued against digitization or, even fewer, that the revolution does not exist, while especially in recent years, a few more people have emphasized its negative effects. That the existence of the digital revolution *cannot* be questioned has been clear at all levels, everywhere in the world, for several decades. That human beings were involved, or rather invested and overwhelmed, by this revolution is just as true. As Malaysian Postal Minister Datuk Leo Moggie said in 1995: "None of us can afford to say that the information revolution does not concern him, be it the head of department, the chief executive officer or the student." And this is continuously repeated when, in affirmative tones, it is argued that digital liberalism is irresistible, as is the compulsive habit of scrolling, clicking, watching, and listening to digital content.

In short, the digital revolution cannot be contested or stopped. And this characteristic seems to recall one of the two meanings of revolution mentioned in Chapter 1, the oldest one in which, as Hannah Arendt writes, the motion of the stars follows "a preordained path" and completing this "revolution" is an irresistible natural phenomenon, "removed from all influence of human power." Hand in hand with this etymology of cyclical return is obviously the idea of historical necessity. It was the French revolutionary Camille Desmoulins who spoke of a *torrent révolutionaire* or "revolutionary torrent," inaugurating the aquatic metaphors associated with the revolution (including the digital one), often described as unstoppable floods, violent tsunamis, and rapidly flowing streams that break down barriers and overwhelm everything in their path. But perhaps the best known quotation about the "irresistible revolution" is from another theorist of revolutions, Alexis de Tocqueville, who in the introduction to the second volume of *Democracy in America* writes about democracy itself: "The whole book which is here offered to the public has been written under the impression of a kind of religious dread produced in the author's mind by the contemplation of so irresistible a revolution, which has advanced for centuries in spite of such amazing obstacles, and which is still proceeding in the midst of the ruins it has made." For de Tocqueville, writing in the nineteenth century, democracy had already been advancing irresistibly for centuries. It seemed like a providential, transcendent, and universal fact; it was proceeding despite the ruins or errors it was leaving in its wake.

To me, it also seems like one of the definitions most applicable to the digital revolution which, like a river in flood, has overwhelmed diverse generations, societies, and cultures in recent decades: it is no accident that expressions like the "digital tsunami," "digital floods," and others have become quite popular. Of course, this flood has sometimes turned into "information overload," as another commonly used expression describes it, and has also caused damage in the process which has (wisely) been kept in the background, creating a kind of "religious dread" that will be discussed in Chapter 4.

Three "laws" often mentioned when it comes to digitization also contribute to making the digital revolution irresistible and therefore unstoppable. The term "law" already suggests that these observations—chiefly advanced by inventor-entrepreneurs—have a sort of scientific legitimacy, are based on empirical data, and are valid until someone disproves them, as is generally the case in scientific research. The fact these so-called laws have been criticized or even refuted hasn't diluted their popularity and they are still venerated and recited almost as if they were indisputable dogmas of the irresistible revolution.

Moore's Law, formulated in 1965 by Gordon Moore of Fairchild Semiconductor and then co-founder of Intel in 1968, predicts exponential growth in the computing power of computers over time. In reality, Moore never formulated this law in the same terms with which it was popularized, nor has he ever thought of it as a real law, nor has the predicted growth of computing power occurred punctually over time, as is often argued. In 1965, Moore noticed that the number of transistors on a microchip had doubled every two years since the first prototype was produced in 1959. This observation, without any predictive value, was then modified over time. In the 1980s it was described as the doubling of the number of transistors every 18 months; in the early 1990s it was interpreted as the doubling of the power of microprocessors every 18 months, and this morphed into stating that the computing power of a computer doubles every 18 months. Many observers have noticed a slowdown in the exponential growth of computer power in recent decades, but these critical remarks have had no impact and "Moore's Law" has maintained its axiomatic validity, driven by Intel's interest in guiding public discourse on the subject.

The second law is known as Metcalfe's Law and it has to do with communication networks. Named after Robert Metcalfe, inventor of Ethernet networks and founder of 3Com, it states that the utility and value of a network are proportional to the square of the number of users. Given its corollaries, it comes as no surprise that this law was internalized and promoted by the Internet world. Connection/connectivity must be raised to the absolute value of the network economy, a concept fully embraced both by political theory, which in fact insists on countering the digital divide, and by the stated slogans and missions of some digital companies such as "Connecting People" (Nokia a few years ago) or "Bring the world closer together" (Facebook since 2017). Consequently, it is necessary to increase the number of users in a network (including a social network site, obviously) as much as possible in order for it to acquire economic value. This includes those who are recalcitrant or who, for technical reasons, cannot access the network and who then risk being increasingly excluded from online and even offline social relationships. In short, Metcalfe's Law states that once it has attracted a large group of users, a network acquires economic and social value, which makes it irreversible and difficult to replace (path dependence theories would say viscous) and even "obliges" users to be part of it. Take WhatsApp at this moment in history: consider, in terms of social relationships, how difficult and costly it is not to use it, but also the economic value it represents for its parent company, Meta. Even Metcalfe's Law, like Moore's, has been criticized and questioned over time, first of all because it is difficult to quantify and verify, but it is now accepted and taken for granted in public discourse—so much so that the economic value of digital companies that offer network services is measured by the number of users.

The last of the three laws that explain the irresistible revolution is Makimoto's Wave, the first version of which was formulated in the early

1990s by the managing director of Sony, Tsugio Makimoto, an important figure in the history of Japanese digitization, so much so that his nickname is Mr. Semiconductor. Makimoto's Wave, also formulated through various graphs, theorizes the existence of 10-year cycles in the development of digital technologies; in particular, a decade of technological development and standardization seems to be followed by another one characterized by the sale and commercialization of that technology. This law evolved over time until 2002, when it settled into a formulation in which Makimoto foresaw various cycles of the digital revolution: a first analog cycle (the contradiction in terms is interesting) between the 1970s and 1980s focused on TV and video recorder technologies; a second digital cycle between the 1980s and 2005 in which the computer took center stage; followed by a third cycle, also digital, characterized by the diffusion of various consumer products and networks, which began in the 1990s and should last until at least until 2030.

The three laws, as a whole, are the "scientific" translation of the deterministic idea that the digital revolution is an irresistible, exponential, and, it could be said, inexorable process: computers will double their power, networks with more users will attract other users and this will increase their value, the revolution will proceed in cycles in which new technologies will inevitably replace old digital devices. These laws can explain the past digital tsunami but, despite getting old, they also maintain an ability to predict future tsunamis. It is indeed clear that these three laws aspire to be predictive and seek to interpret, or in some cases to shape, the future of digitalization, as truly self-fulfilling prophecies. But the rhetorical role of the future in the digital revolution deserves a separate section.

3.4 A Revolution for the Future

Like all ideologies and most revolutions, the digital revolution is also full of promise. Moreover, the ideal historical time for the digital revolution is, of course, the future. This is also because in the present it is always described as incomplete, as a constantly evolving process, as a transformation which does not and cannot have a predefined end. But which future are we talking about?

The digital revolution is linked with two iterations of the future: either a future that is already here or a near future that is about to come to fruition and that precisely because it is so imminent is easily predictable.

A quotation attributed to the best-known cyberpunk writer, William Gibson, reads: "The future is already here—It's just not very evenly distributed." The first part of this quotation reminds us that the digital future is already here, and that in fact the digital revolution (and now increasingly the transformation) is the future that is taking place before our eyes. This future-present has also become part of many companies' mission statements. To offer just a few examples, we can start with Toshiba: "Committed to people, Committed to the Future," or one of Microsoft's most famous and durable mottos: "The future is now," while "The future is today" was adopted by the Japanese semiconductor firm Gigaphoton in 2019. Also think about the promotional campaigns for the Metaverse: the project was launched by Mark Zuckerberg in person in late 2021 and the ads proposing future applications of the virtual world emerged especially in 2022 all with the same claim: "The metaverse may be virtual, but the impact will be real." For digital companies, in particular, squeezing the story of the future of the digital revolution into the present is strategically important. By doing so, they describe themselves as innovative companies today and imply that they will remain so tomorrow. They leave no room for uncertainty because they already know what the future will be like, they allow you to travel through time (for example, in the Metaverse ad, a group of students travels back to the Roman Empire) and learn about the future while remaining firmly situated in the present. We must have patience if this future-present is full of inequalities, as the second part of Gibson's quotation reminds us; they'll soon be rectified.

The second type of future, which is often the theme of the narratives that accompany the digital revolution, will actually take place in a time that is yet to come, even if it is in the near future so as not to risk making incorrect predictions that would generate uncertainty. This future is especially expressed through predictions or even prophecies about the digital revolution, which are, not surprisingly, often optimistic. They range from predicting that the world will improve thanks to the digital revolution, to the fact that the revolution itself will be a success which will lead to linear economic growth, to the prediction that I think is more interesting: that the digital revolution *will inevitably continue in the future*.

These predictions almost always bet on the linear growth of users who access the network, who purchase mobile telephone contracts, who own computers, and who subscribe to social media. In the coming years—usually five or ten—the revolution will be even more disruptive, it will finally be fully realized and will allow us to live in new social realities, to cite some of the most famous predictions. The moment in which the prediction of continuity was most strategic was also the most delicate moment in the history of the revolution: the collapse of dot.com stocks in 2000–2001. In fact, in those uncertain years, companies and commentators were quick to declare that the revolution would not end and that there would be new positive cycles. At the Birmingham IT forum in 2003, Graham Whitehead admonished the audience: "We are about to enter the Information Revolution—and if you think that we are already there you have a rude awakening awaiting for you in the next three to five years." Also in 2003, the president and CEO of Samsung Electronics, Choi Gee-sung, argued that "the digital revolution is just beginning" and, in 2006, that "the consumer electronics market will see a 'digital boom' along with the opening of a 'digital golden age.' Simply put, another digital revolution is coming." Ernst Malmsten, the Swedish co-founder of the fashion site Boo.com (which later failed), drew the funniest parallel in 2002 when he said: "I believe the dot.com revolution has now gone into hibernation, like the Vikings who stayed at home during wintertime, building their longships and preparing for the ice to break up." After a harsh and complex winter, like the bursting of the Internet bubble, we really wanted to believe that the spring sun would shine again on the digital revolution. Positive predictions are useful, especially when the digital revolution slows down and billions of dollars and jobs are lost. It is no accident that when the Big Tech crisis of 2022 hit—with thousands of people around the world fired from their jobs, with the collapse of several cryptocurrencies like BitCoin, and with the Nasdaq losing about one-third of its value in a few months—it led to a new wave of predictions. For example, supporting the idea of the metaverse, Mark Zuckerberg claims that "[r]emote work is going to be a bigger part of the future. I think within five to 10 years, probably about half the company is going to be remote." Or Nancy Gohring, research director of the IDC Future of Digital Innovation, acknowledged "the multitude of current headwinds ranging from inflation to war," but reassured the market claiming that "winners and losers in each sector will be determined by their ability to deliver digital innovation at scale—ethically, sustainably, and repeatedly."

But who develops these predictions and does the "forecasting," even to the point of making what can rightly be called prophecies? Forecasting has become a common "genre" among experts and commentators on digital issues, and these include politicians, international and non-profit organizations, academics, journalists, and the CEOs of large companies who often organize conferences to prophesy about the future of the revolution, companies themselves, or specific departments (for example, since 2010 Adobe has been publishing their annual *Digital Trends* reports, which often feature forecasts), as well as specialized consulting businesses such as McKinsey, Deloitte, and Gartner. All these stakeholders in the digital revolution have tried their hand at the art of prediction which, to tell the truth, increasingly assumes the trappings of a real science or, better, of a rational business strategy to adopt in the present for some fundamental decisions. In fact, forecasting is not only a way to anticipate the future, which is full of uncertainties and competing alternative visions, but it is also (and above all) a way to shape and build that future. The decisions made today, in the name of a tomorrow that is necessarily considered digital, strongly influence the future itself because not only do they conflate consumers' economic and emotional investments (which includes even just learning to use a new digital device because *that is the future*), but they also guide political or business interests and cultures that thus focus on one of the many possible horizons, which obviously entails discarding others.

Among the best-known manufacturers of digital predictions and prophecies, *Wired* plays a special role. Indeed, in the 25th anniversary issue of the magazine, published in October of 2018, senior editor Adam Rogers recalled:

> The main justification for our claim that the digital revolution would reshape the world had been brash confidence and what our founding editor in chief [Rossetto] calls "militant optimism." And then, somewhere around 2003, it turned out to be true . . . As an editor, I used to tell writers: WIRED doesn't make predictions. Write about things that are happening right now, today—the unevenly distributed upwellings of the future here in the present, to borrow from William Gibson. Our stories sound like science fiction. But they're true.

This quotation contains many elements crucial to understanding the relationship between the digital revolution, forecasts, and the future. First and foremost, the digital revolution is a type of secular religion that requires trust in the future and optimism. And this optimism must be militant; in other words, it must help to disseminate the "word" of the revolution. *Wired* is a print evangelist of the revolution and, as *The New York Times* already noted in 1995, its genius "is that it makes the Digital Revolution a self-fulfilling prophecy, both illuminating this new subculture and promoting it—thus creating new demand for digital tools, digital toys, digital attitudes." In short, the prophecies of the digital revolution are a type of business promotion and also involve the creation of a mythical universe.

Secondly, these predictions are only predictions for "ordinary" people, while the visionaries of *Wired* (and, by extension, for all the prophets of digitalization who read it) can describe the present-future they already see—after all, what kind of visionaries would they be if they didn't? However, only a few people can truly grasp this present-future: the elite of the digital revolution. In this regard, Gibson's passage quoted above is interesting, but with a fresh interpretation: inequality and a lack of homogeneity in the distribution of the future gives way to the idea that the word of revolution can reach or reveal itself only to a select few, the happy few, who today, already know what's next.

And "What's Next?" is probably one of the most symbolic and repeated mantras of the digital revolution in its desperate attempt to understand the future. This is also the question that observers of the digital revolution obsessively ask themselves. In general, the answer is new devices and digital services that are more innovative than the previous editions (which, however, were only recently described as super-innovative). Next there may be a new phase, usually an acceleration, of the revolutionary or transformative process. Furthermore, the "next" must also be known as soon as possible in order to understand where the revolution is going and what to invest in. The way in which the financial dimension of the digital revolution is combined with the future is central: it should not be forgotten, in fact, that this is a revolution paid for and driven by investors, and by stocks on the stock market, which are, de facto, a gamble, and which try to predict future profitability also on the basis of past performance. When we turn the issue around, it can be argued that ensuring a given future for the digital revolution, and all the actors that revolve around it, is equivalent to attracting investments and stock equity in the present. The future generates revenues.

3.5 A Permanent Revolution

The concept of "permanent revolution" was introduced by various classics of nineteenth-century sociology such as Karl Marx and Pierre-Joseph Proudhon and then popularized by Leon Trotsky during the Russian Revolution. Despite the different nuances of meaning, Marx and Proudhon understood that in order to implement a profound transformation of socio-economic relationships and cultural traditions, not only would it take a long time, it would also be necessary to think of the revolutionary process itself as a permanent phenomenon, never fully concluded and capable of constantly renewing itself. On the other hand, Trotsky believed that revolutionary times should be compressed, that the (socialist) revolution should break out in the present and that society would not find peace until it had been fully realized.

These meanings are present in (and it must also be said that they characterize) the digital revolution, which has a constant need to be regenerated, to find new impetus even in old slogans, and to recast the past—in an often propagandistic function—in order to stay alive. It is not blasphemous to compare Jeff Bezos to Karl Marx when, in his last letter to shareholders before leaving the leadership of Amazon in 2020, he invited them to continue in the constant and peaceful revolution that has always characterized the company, to remain unique and constantly changing to stand out in the contemporary digital universe.

And, in fact, the digital revolution and transformation need to be self-described and narrated as "permanent" for at least two strategic reasons. First, the digital revolution is permanent because in fact it continues to occur, potentially infinitely, once again calling to mind the first meaning of revolution as a continuous and perpetual motion of heavenly bodies. Because it is continuously renewing itself, the revolution has never been completely realized, and it always poses new challenges. On the contrary, the revolution will remain permanent until its goals are achieved and, precisely for this reason, it is important that they are never achieved because if they were, the revolutionary drive, as well as business, would be finished. However, in order to remain permanent, the digital revolution paradoxically needs to insist on slogans touting rupture and subversion with respect to a past that, with the passage of time, doesn't seem to age and, on the contrary, has been exported and readapted to various regions of the world. The rhetoric is often similar (among others, the best is yet to come, we are in a new phase of the revolution, the revolution is winning), but contexts and historical moments are changing. This need to regenerate disguises an economic and political need: to remain a driving ideology, the digital revolution must continue to inspire, to sell persuasive products, to be perceived as just starting, as unfinished, as a harbinger of new business opportunities. But above all, it needs to offer visions for the future through the words of specialists in this genre, such as Bill Gates, who in 2008 declared:

> People often ask me if we're nearing the end of the digital revolution . . . I believe the opposite is true. In many ways, the incredible advances of the past few decades have really just laid the foundation for much more profound change . . .Together, hardware and software will be the catalyst for advances during the next 10 years that will far exceed the changes of the last 30 years.

And then in 2023, joining the hype of the artificial intelligence, he claimed in his Twitter profile linked to one of his Gates Notes:

> The development of AI is as fundamental as the creation of the microprocessor, the personal computer, the Internet, and the mobile phone. It will change the way people work, learn, travel, get health care, and communicate with each other. . . I think in the next five to 10 years, AI-driven software will finally deliver on the promise of revolutionizing the way people teach and learn.

Gates has made dozens of short- or medium-term prophecies like these, all characterized by the idea that the revolution will continue in the future. In these short quotations then, there are a variety of digital revolution mantras, like the ones that have been described in this chapter: there is the idea of disruptive changes to come, the irresistibility and

"Whoa! We sure blew that Prediction!"

Figure 9 Bill Gates' prophecies and prophecies about Bill Gates.
© CartoonStock.com, cartoonist Mark Ishikawa, uploaded September 1, 2011.

linearity of progress, betting on its future persistence and on the fact that the best is yet come. Both in 2008 and 2023, the revolution seemed to be *just beginning*, the following years were to be the most significant and most transformative, not only had the revolutionary drive not been exhausted, on the contrary, it seemed destined to persist and to grow.

In recent years, the prophetic role of Bill Gates seems to have been taken over by Elon Musk: despite some huge mistakes (like his prediction regarding the growth of cryptocurrencies in 2021 or continuous failed forecasting regarding the conquest of space), Musk's role as a prophet is certified by the press and specialized websites, making him the new idol of prophecies. Musk is generally more pessimistic than Bill Gates and, for example, in 2016 he predicted that "AI will be capable of performing any task as well or better than humans—otherwise known as high-level machine intelligence—by 2060 and will overtake all human jobs by 2136"; however, the persistence of digital revolution or transformation is never questioned but, again, it is seen as the most relevant driver for human change.

To fully understand this first dimension of permanence, it is also useful to resort to Hannah Arendt's concept of *revolutionary dictatorship*, according to which the most common outcome of revolutions is the establishment of a dictatorship whose objective is to "drive on and intensify the outcome of modern revolution" itself. And the story of the digital revolution is a typically dictatorial one: the main objective seems to be perpetrating the digital revolution itself, extending its lifespan and significance as much as possible and, consequently, continuing to ensure that it's profitable for professional revolutionaries.

Then there is another meaning of permanent, which concerns leaving an indelible mark—permanent precisely from the Latin *manere* in the sense of lasting and remaining—that the digital revolution leaves behind. We cannot escape the changes made by the digital revolution. It is impossible to imagine returning to living in an analog or pre-digital world, whatever that means, and even if it were possible, because the effects of the revolution are indelible and chronic. It has often been discussed, for example, whether at the end of the COVID-19 pandemic we will return to the "way things were before," and this discussion also involves the digital. According to most observers, the pandemic has accelerated or even made the digital transformation a chronic condition and, in addition, phenomena such as smart working, the need for broadband, and the acquisition of digital skills are considered part of an irreversible future. In short, the post-pandemic future will be even more digital than the past—you can bet on it.

Figure 10 COVID-19 as the main leader of digital transformation.
© and cartoonist Virpi Oinonen, 2020. Appeared in https://www.businessillustrator. com/what-is-digital-transformation-cartoon-infographic/ and other online platforms.

The digital revolution has been described as disruptive, total, irresistible, inevitably associated with the future, and permanent. Many of these mantras are also characteristic of previous political-social revolutions and, once again, by adopting "classic" revolutionary languages and stereotypes, the digital revolution was narrated and conceived as the heir to a long revolutionary tradition. Of course, the digital revolution also differs from many past revolutions and the way in which they were studied and thought out. First of all, the desire is for it to be a non-conflictual revolution that brings everyone together, so much so that few have argued against the idea that we are living in an era of digital revolution. Furthermore, at least in recent decades, its purpose is not to overthrow a pre-established political order and establish a new one; if anything, the intention is to initiate a new economic order thanks to which more products or ideas can be sold and thus generate wealth. Its protagonists are not politicians, military generals, or entire social classes bent on overthrowing the established power: the leading figures of the digital revolution are entrepreneurs who have little of the heroic about them and are actually rather uncool and nerdy. The weapons of the digital revolution almost never kill but they invite you to buy products or compulsively absorb consumers' attention. Finally, the revolution is played out in places that are not town squares, institutions that are the seats of power, or the battlefields, but rather in the company headquarters, the stores, and the high-tech trade fairs where millions of consumers go to see the revolution up close. This revolutionary pantheon, indispensable for creating a religious and ideological aura surrounding the digital revolution, will be analyzed in Chapter 4 and the Conclusion.

Endnotes

It seems that the initial question about the "revolutionary" nature of the digital revolution would be answered by Zysman and Newman 2006 but, however, in reality this book deals with something else. To describe the slogans adopted by the digital revolution, I decided to use the term *mantra*, whose original meaning comes from the Hindu religion and indicates a "repeated word or sound" used as a meditation aid (from Sanskrit, literally "a thought, thought behind speech or action," see https://premium.oxforddictionaries.com/definition/english/mantra).

Section 3.1 mentions a passage by Karl Griewank taken from *Der neuzeitliche Revolutionsbegriff*, quoted and translated into English in Hobsbawm, 1986, p. 9. The other two typical characteristics of revolutions according to Griewank are "a social content, which appears in the movement of groups and masses, and generally also in actions of open resistance by these," and "the intellectual form of a programmatic idea or ideology, which sets up positive objectives aiming at renovation, further development or the progress of humanity." These two definitions can also be adapted to the digital revolution in varying degrees and the third is, in fact, one of the fundamental themes of this book. The literature on economic transformations induced by the digital revolution is vast and some authors have already been mentioned in Chapter 1. The text mentions Gates, 1995, in particular chapter 8 of that book which focuses on "friction-free capitalism." On the parallels between the digital revolution and the discovery of fire, see Attali, 1992; Barlow, 1995; Rossetto, 1993. On AI as a discovery more important than fire, see Knowles, 2021. The comparison with fire has also spread to academic work, see for example Fischer, 2006, where in the blurb it states that the author "believes that the digital revolution is a definitive moment in human history, as important as the discovery of fire. Deceptively quiet, it is invasive, radical, and affects all aspects of human activity."

On the meanings of total in Section 3.2, see *The Compact Edition of the Oxford English Dictionary*, 1976, p. 176. The four similar quotations, made decades apart, on the potential transformation of human nature are taken from Friedrich, 1983; from the January 1998 issue of *Wired*; from Zuckerberg, 2014; and from the World Economic Forum, 2017. On the negative transformation of humans due to the digital revolution see O'Lemmon, 2022. In this paper, there is also a comparison between the negative effects of the agricultural and the digital revolutions. Among the classics of communist revolutionary thought and, more generally, on the historical evolution of revolutionary thought, see the excellent work by Lenk, 1973 and, in English, Cohan, 1975. As an example of RAND reports on the digital revolution and transformation, see Hundley et al., 2003. For ITU, I am referring to the article by Deputy Secretary General Malcolm Johnson, 2019. In *Being Digital* by Negroponte, 1995, the quotation about rock music can be found on p. 204. On the collective singular of "the revolution," see Koselleck, 2004, pp. 50–53. Larry Page's quotation about digital technology being capable of solving humanity's problems is taken from Waters, 2014. Solutionism is quite popular topic in the work of the digital theorist Evgeny Morozov, 2014, where he also mentions that the Internet "It has also given rise to a new set of beliefs—what I call 'Internet-centrism'—the chief of which is the firm conviction that we are living through unique, revolutionary times, in which the previous truths no longer hold, everything is undergoing profound change and the need to 'fix things' runs as high as ever" (pp. 15–16). Finally, the achievement of the collective good and the reduction of poverty as two of the main goals of revolutions are mentioned Arendt, 1990, p. 60.

On the meanings of irresistible in Section 3.3, see https://premium.oxforddictionaries.com/definition/english/irresistible and https://premium.oxforddictionaries.com/definition/english-thesaurus/irresistible. In relation to the attractiveness of digital devices, see Alter, 2017; Sadin, 2016. Minister Moogie's statement can be found in Kushairi, 1995. Hannah Arendt's quotation is taken from the often cited *On Revolution* (1990, p. 47), as is the expression "*torrent révolutionnaire*" attributed to Desmoulins (p. 48). The quotation from de Tocqueville, 2004 is at p. 7. On the metaphor of flood, see Cortada, 2012. On Moore's Law, the literature is vast, but for a critical look at the law and its mythologies, see Tuomi, 2002. On Metcalfe's Law, see Briscoe et al., 2016. Makimoto's Wave and related graphics can be found in Merrit, 2016, chapter 2. On the laws of digitalization as self-fulfilling prophecies, see Gabrys, 2011, p. 116. Finally, I am not the first one to combine different computer laws, see for example these and other laws mentioned in a special issue edited by Getov, 2013.

Section 3.4 deals with the future. What is currently known as "future studies" have extensive literature that is not easily summarized, but I think Barbrook, 2007 is a text that contains many of the elements collected here: the ideological and imaginary dimension of the future, its relationship with the present and of course digitalization. In Italian there is the impressive work of Jedlowski, 2017. Gibson's quote about the future at the beginning of the paragraph is uncertain and the writer himself does not remember having uttered it (see "The Future has Arrived," 2021). Cyberpunk is one of the great surprises in this book: I would have expected to find many quotes about the digital revolution, on the transformative dimension of digitalization, and on disruptive change. Instead, after consulting dozens of books and anthologies, I can say that cyberpunk sources don't really deal with the "digital revolution" (or at least almost never use this expression), but have highlighted a more critical dimension, of slow and profound change. Going back to the future, some business mottos are also mentioned: those of Toshiba and Microsoft are known, while the reasons that have led Gigaphoton to change its own are found in "Gigaphoton Announces New Corporate Slogan," 2019. The Metaverse ads can be easily found on YouTube. The discourses around the Metaverse are full of the ideological rhetoric of revolution, which is also embraced by publications (see Ball, 2022).

On forecasts in the field of communication technologies, see chapter 2 of Carey and Elton, 2018. On the repetitiveness of predictions/prophecies in the digital revolution, Krzywdzinski et al., 2018. Among the forecasts for the continuation of the revolution after the bursting of the dot.com stocks' bubble, Graham Whitehead's quotation is taken from "Information Revolution Is Only Just Starting," 2003; those by Choi Gee-sung are found in Ghosh, 2003 and in Dong-woo, 2006. Ernst Malmsten's quotation about the revolution's "lethargy" is taken from "Business Monthly," 2002. Mark Zuckerberg made the prediction on the future of work in Robinson, 2022. Nancy Gohring's words are taken from her 2022 report.

On some consulting firms, such as Gartner, that influence the digital future, see Pollock and Williams, 2016. On *Wired* and prophecies, the quote comes from Rogers, 2018, and Keegan, 1995. One of the most interesting articles on *Wired*'s "futurological" approach, which also includes some self-criticism, is Karpf, 2018.

For the meanings of permanent in Section 3.5, see https://premium. oxforddictionaries.com/definition/english/permanent and https:// www.etymonline.com/word/permanent#etymonline_v_12722. On the classics of the permanent revolution, I refer once again to Lenk, 1973, in particular chapter 6. The parallel between Marx and Bezos is proposed in Collard, 2021. Bill Gates' quotations can be found in an article he authored in 2008, p. 14, and in his Twitter profile (Gates,

2023). A short history of the prophecies of Elon Musk is given by Chirinos, 2022, from which the quotation is taken. Arendt writes: "Not constitutions, the end product and also the end of revolutions, but revolutionary dictatorships, designed to drive on and intensify the revolutionary movement, have thus far been the more familiar outcome of modern revolution—unless the revolution was defeated and succeeded by some kind of restoration" (1990, p. 159).

4

Believing in the Revolution
A Contemporary Quasi-Religion

The digital revolution is almost like a religion, not so much because it has transformed religious practices, as it has every other area of human activity, but because it has often been described as a belief system, an act of faith, and a moment of enlightenment that has affected humanity. It is a religion of our era or, at least, it is often described using sacred and religious metaphors: the protagonists of the digital revolution are gurus, messiahs, and tireless evangelists; the places where the revolution is happening are the meccas toward which we should turn; digital objects are sacred relics and transport us to transcendent realities. And then there are even heretics and infidels who have taken the liberty of opposing the positive idea of digitization, but do not question its revolutionary character.

But there is something more: the religious aspect of the digital revolution is added to the ideological one. In 1984, historian of technology Langdon Winner used the term *mythinformation* to describe

> the almost religious conviction that a widespread adoption of computers and communications systems, along with broad access to electronic information, will automatically produce a better world for humanity. It is a peculiar form of enthusiasm that characterizes social fashions of the latter decades of the twentieth century. Many people who have grown cynical or discouraged about other aspects of social life are completely enthralled by the supposed redemptive qualities of computers and telecommunications . . . it is an expressive contemporary ideology.

> I use the term *ideology* here in a sense common in social science: a set of
> beliefs that expresses the needs and aspirations of a group, class, culture,
> or subculture.

In 2004 in perhaps his most significant book dedicated to the concept of
the digital sublime, Vincent Mosco highlighted both the uncritical and
ideological enthusiasm generated by digitization and the quasi-religious
tone of the narratives that have accompanied it over time. Just as in the
Romantic concept of the sublime, in which the traveler was seized by
a mixture of wonder and fear when observing the mountains at high
altitudes, so the modern travelers of the digital age are both fascinated
and frightened by the digital. The concepts of mythinformation and
the digital sublime have at least three characteristics which can be
traced in this book: they are quasi-religious convictions that supersede
or add to other belief systems or faiths; they are ideological in nature
because they express beliefs, needs, aspirations, and ultimately overall
world views; and, finally, they represent a type of fervor for and blind
trust in technology. These three qualities can also be applied to the
idea of a digital revolution.

One of the key verbs linking digital revolution, religion, and ideology is *belief*. Believers place their trust in a person, idea, or doctrine
and don't question its truthfulness or rationality. This suspension of
disbelief, or rather blind trust, has also characterized the advent of the
digital. Belief in digital media, Simone Natale and Diana Pasulka write,
is articulated in at least four categories: that digital devices work and
are therefore reliable for carrying out everyday tasks; that they are profoundly different from previous technologies; that they change or are
about to change human culture and societies irrevocably; and, finally,
that they can even transform human nature, creating hybrids between
humans and machines, and defeating death, as mentioned previously. Believing in the digital *revolution*, we might add, means trusting in
a clean break with the past (after all, even in some religions there is
a periodization based on a *before* and *after*), hoping in the transformative power of digital devices, *trusting in* the words and visions of the

gurus who have been preaching the change that is underway for several decades.

This quasi-religious aspect emerges clearly in some of the narratives surrounding the digital revolution. From blogs urging readers to *keep faith* with the revolution despite several years having passed since its "birth" (whenever that may have been), to the well-known parallelism between Google and God, which has even spawned a Reformed Church and a denomination with ad hoc prayers and commandments.

A religion founded and recognized in Sweden is the church of *kopimism*, a belief system that is based on the adoration of the download, of file sharing, and above all of the concept of *copy and paste* (so much so that CTRL+C and CTRL+V are considered sacred symbols). Then there are, for example, company names, such as Oracle or Lenovo (which is a portmanteau combining the company's original name, Legend, with a Latin form of the word "new"), or services such as the Cloud or Big Data, which immediately call to mind the religious, supernatural, and extraterrestrial dimension of the revolution. Computers, the Internet, the Web, smartphones, and many other devices and services have been described as blessings from heaven, new divine tools, supernatural revelations. In short, digitization has often used religious or para-religious language to describe itself or be described. In this chapter, for the sake of clarity and simplicity, we will identify those protagonists, objects, and places associated with the digital *revolution* that have most frequently or most significantly taken on quasi-religious connotations. The intent is to understand how the digital revolution, through its actors and actants, has taken on a mystical and transcendent character, forged by formulas, slogans, or even commandments that have been repeated over time.

4.1 Patriarchs and Patron Saints

In the beginning was the digital. Virtually all religions have founding figures, spiritual fathers (much less frequently mothers) who collect the teachings of the deity and pass them on to their followers.

In the Jewish and Christian religions, for example, this role is played by the Old Testament patriarchs: from Adam to Noah, from Methuselah to Abraham, and many others. Patriarchs are endowed with great longevity; they are vested with experience, authority, and wisdom; they merit respect and obedience; they are heads of families or communities to whom special powers are conferred along with a leadership role.

Even the digital revolution has its patriarchs. In fact, there are people who are cited again and again because they understood the transformative reach of digital from the very beginning. Typically, they are recognized as such over time through the investiture of certain individuals who have every interest in generating a pantheon of founding fathers. Google frequently dedicates Doodles to them; newspapers and magazines celebrate the birthdays, anniversaries, or deaths of the heroes of the revolution in panegyric obituaries; the same digital corporations exalt their own present by constructing, often artfully, a glorious past full of precursors and visionaries.

This pantheon is flexible and variable in the sense that some patriarchs can fall into disgrace while others re-emerge from oblivion; so, drawing up a complete list is impossible, although some categories and professions are privileged over others. For example, there are mathematicians who, especially since the nineteenth century but in some cases even earlier, have retrospectively laid the foundations for the creation of modern computers. George Boole with Boolean algebra, now recognized as central to the analysis and synthesis of switching networks, "enabled revolutionary thinking in not just logic and math, but also engineering and computer science." Charles Babbage and Ada Lovelace (a rare woman "visionary of the computing age") built what is considered one of the first mechanical calculators, the analytical engine, wrote one of the most important algorithms in history, and anticipated the importance of computing machines beyond the realm of mathematics. Claude Shannon, who worked at Bell Labs (the research department of America's leading twentieth-century telephone company, American Telephone & Telegraph), wrote a well-known paper on information theory that "led to revolutionary changes in the storage and transmission of data." Norbert Wiener and cyberneticians in general have explored the communicative interactions between man and machine (after all, religion comes from the Latin *religare*, to bind). However, the most famous mathematical patriarch is Alan Turing, who conceived what is known as the Turing Machine (actually not a mechanical device, as is often thought, but a programming language that today we would call software), devised the Turing Test to assess a computer's ability to mimic human behavior and, moreover, during the Second World War he was involved in building the Colossus computer, which was able to decipher encrypted Nazi communications. Turing has become a "mythic" celebrity of the digital revolution: his test was glorified in 1982 in *Blade Runner* (one of the cult movies of the revolution) and another very popular film from 2014, *The Imitation Game*, told his tragic human story. Dozens of books and hagiographic articles have been written about him, and statues of him were even erected in Bletchley Park, north of London, and in Sackville Gardens, in Manchester—after all, monuments are erected to commemorate heroes of revolutions. Turing is considered a genius, his work a forerunner of contemporary digitization, who "laid the foundations for modern computers and the information technology revolution."

Then there are patriarch-entrepreneurs, related to the world of early computers, some of whom have already been mentioned. Gordon Moore predicted the irresistibility of the revolution with his law. Andrew Grove, also of Intel, and chip "visionary" Paul Allen, cofounder of Microsoft, introduced new technologies. Of Allen it is said, for example, that he "conceived many of the ideas which became cornerstones of the digital revolution." Doug Engelbart, who was one of the first developers of the computer mouse and some graphical interfaces, "foresaw the information revolution." In time, Steve Jobs, Bill Gates, Elon Musk, Jeff Bezos, Mark Zuckerberg, Larry Page, Sergey Brin, Jack Ma, Ma Huateng, and many other digital heroes of the last decades, who will be found in Section 4.2 of this chapter, are likely to turn into patriarchs and founding fathers and, indeed, in the case of some the rise has already begun. After all, the fluidity and variability of the digital revolution is typical: as already mentioned in Chapter 1, the revolution needs to continually regenerate itself and therefore it also needs new figures to display in its pantheon.

In addition to mathematicians and entrepreneurs, some patriarchs of the revolution are found among academics and scholars. As already mentioned, some political scientists and economists, especially between the 1960s and 1970s, theorized the advent of new social structures clearly distinct from those of the past and in which new communication technologies (digital and otherwise) would allow humanity to become part of information society. In addition to Daniel Bell, Zbigniew Brezinski, Alvin Toffler, Simon Nora, and Alain Minc, the list of those who first predicted, foresaw, and interpreted the impending revolution often includes Fritz Machlup, Peter Drucker, Marc Uri Porat, and Yoneji Masuda. The common trait of these patriarch-academics is their ability to be heard by politicians: not only because much of their research was the result of reports commissioned by governments but also because their theories were translated into policy actions that aimed to transform the digital revolution from a theoretical construct into reality.

Hackers, too, have often translated countercultural information utopias into reality, or at least virtual realities. *Wired* journalist Steven Levy, in a book published in 1984 and then in a new edition in 2010, has traced the history of the hacker community since the 1950s, identifying some founding fathers of the movement but also tracing the evolution of the term "hacker." The connotation of this term hovers between a positive libertarian and a negative intrusive one—the latter is currently predominant in the common imagination, but in its original sense the term was intended to designate "only" a group of experts in computers and programming languages. In the preface to the volume Levy wrote, in quasi-religious tones:

> Hackers like Richard Greenblatt, Bill Gosper, Lee Felsenstein, and John Harris are the spirit and soul of computing itself. I believe their story—their vision, their intimacy with the machine itself, their

experiences inside their peculiar world, and their sometimes dramatic, sometimes absurd "interfaces" with the outside world—is the real story of the computer revolution.

The patriarchs of hacking are the spirit and soul of the digital revolution, they embody its *true* history, they create parallel and mystical universes, they inevitably clash with the "real" world. It sounds more like a description of saints than computer geeks, but it precisely encapsulates all the mythology and idealism/ideology with which these communities have been viewed in recent decades. A mythology that has certainly been reinforced by small gimmicks, including, for example, the "hacker" look, and especially by certain pop culture products, such as television series. Perhaps the most significant example is *Mr. Robot* (whose second season in 2016 was promoted with a trailer titled *The Revolution Continues*), which features a sociopathic computer stalker (Elliot) and a mysterious character (Mr. Robot) who aims to expose the interests of the world's powerful and subvert the social order—in short, really launch the revolution starting from the digital.

However, there is a patriarch more relevant than all those mentioned so far. A person who was named by *Wired* as the "patron saint" of the magazine and of the digital revolution itself, Herbert Marshall McLuhan. The Canadian scholar was already considered a "saint" in 1993 for having introduced visionary concepts such as the global village and for having anticipated the idea of cyberspace, but he was officially invested with the office of patron saint between January and February 1996, when *Wired* published two articles on McLuhan. The first one was an imaginary and posthumous interview (McLuhan died in 1980) with the author himself, almost like a séance served up in digital sauce. In fact, in 1995 someone calling himself Marshall McLuhan began posting on one of the magazine's popular mailing lists and the journalist Gary Wolf began a dialog with the mysterious poster, culminating in an interview. Wolf was convinced, without ever being certain, that he had interacted with a bot programmed to respond as McLuhan would have done.

But why was McLuhan chosen as the patron saint of *Wired*? Perhaps because, by the magazine editors' explicit admission, his 1964 *Understanding Media* was a source of inspiration for the visual manifesto that appeared in the second issue. Moreover, in McLuhan's best-known book the word *revolution* appears 51 times: from references to the French and Industrial Revolutions to the socio-cultural revolutions caused by the various media (analog and, to a lesser extent, digital) throughout history. To be honest, the text also argues the opposite thesis, namely that new communication tools emerge in continuity and as a gradual evolution of the old ones (think of the metaphor of the "horseless carriage" to indicate cars and many other similar examples). But the main reasons for his canonization are spelled out in the second article, the one from February 1996, entitled *The Wisdom of Saint Marshall, the Holy Fool*. According to *Wired*, McLuhan was in fact a beacon for interpreting the nascent digital revolution in the midst of its "tumult," a spiritual guide capable of issuing warnings and admonitions, like "quaint aphoristical exhortations and eschatological prophecies of the early church," and who, precisely because of this ability, was subjected to "a form of martyrdom" due to the fact that "respectable folk turned up their noses at his odor of sanctity." But *Wired* aimed to re-establish the facts, eulogizing the man who, in the meantime, had become a classic figurehead for communication studies and who therefore "lives on, even composing books after his death, as electronic culture's immortal saint." In short, the digital revolution needs patriarchs and patron saints, but also immortal martyrs.

4.2 Prophets, Evangelists, Messiahs, and Gurus

Prophets, evangelists, messiahs, and gurus are not interchangeable terms because they belong to different belief systems and religious faiths and, in fact, are figures with different tasks and functions. But in the social discourse supporting the digital revolution, these terms have often been used synonymously to refer to all those characters, usually male, white, and mostly from the United States or otherwise influenced by US culture and lifestyle, who made and popularized the revolution. The common traits of these characters are essentially three. First, the tireless activity of propagandizing, evangelizing, and exalting the ideas of the digital revolution to an audience of disciples in constant need of motivators and spiritual teachers to follow. Secondly, since they're the ones who made the revolution or understood it before and better than others, they were able to become spokesmen and interpreters of the revolution, to convey its deep meanings and dogmas. Finally, these figures have often taken on the burden of predicting the future, of speaking before things happen, of expressing themselves in prophecies so vague as to be self-fulfilling or adaptable to various situations. In short, they are the professional revolutionaries, those who have breathed fire into the digital revolution and have unmasked the plots and counter-revolutionaries who opposed it.

All of these terms have been so thoroughly assimilated into the public discourse that a growing number of professionals are self-labeling themselves on LinkedIn as "digital evangelists" or "digital transformation evangelists" or even "digital gurus." But for every self-respecting religion, there are also new terms coined ad hoc, which quickly enter the revolutionary jargon, and one of the most significant is *digerati* (more rarely *digirati*). First used in a January 1992 article in *The New York Times* by John Markoff, one of the evangelist-journalists, the term is intended to designate a new elite of "digital literati" (hence the portmanteau), or influential thinkers in the world of digitization, particularly in Silicon Valley, with years of experience behind them, who "promoted a vision of digital technology and the Internet as a transformational element in society." In short, the digerati are digital VIPs, often entrepreneurs with a flock of acolytes and imitators who follow them as if they were gurus.

But who are we talking about? Because there are so many aspirants to these titles, for the sake of simplicity here they have been divided into four main categories.

The first includes politicians. There is virtually no government in the world that hasn't implemented public policies and investments in support of the information revolution or the digital revolution and now the digital transformation. By endorsing digitization, often in good faith and without contradiction, global politics has acted as a powerful evangelist. Just read or listen to the speeches of Obama, Trump, Xi Jinping, Modi, Putin, and many other world leaders. Even in North Korea, Kim Jong-un seems to have embraced the ideology of the digital revolution and wants to promote it in his country. Politicians and policy-makers are both genuinely convinced evangelists and good prophets: because they have to justify and account for their strategic choices and the use of public money to enhance digital infrastructure, their speeches are not only a bet on the future but also a way to direct it, to shape it.

A second category is that of journalists and others who write about digital issues. In the bibliographical references of this book you will find the names of many such writers; their prototype and progenitor is Louis Rossetto, true power behind the throne and evangelist of *Wired*, followed by many other journalists and editors of the magazine, including Chris Anderson. To give three more examples, we must also remember John Markoff, who for years has chronicled the revolution for *The New York Times*, including critical remarks; Walter Isaacson, who has worked for several newspapers and has become the "official" biographer of the revolution having written a bestseller on the life of Steve Jobs and an apologia on the revolution itself; and Henry Blodget, founder of *Business Insider*, a website in five languages that is very popular among the acolytes of the revolution, who has often predicted the advent of digital revolutions of all kinds around the world and has reassured the continuation of digital transformation even in the most difficult months of the COVID-19 pandemic. More generally, one can observe that in practically every country in the world, a relatively small group (or sect?) of much-read journalists has specialized in spreading the revolutionary ideology and in carrying out an even more relevant and persistent function than that of the politicians: they reassure readers, on a daily basis, that the revolution not only exists but that it continues in spite of, or even thanks to, global events such as the climate crisis, pandemics, or wars. In short, journalists are the parish priests of the revolution who have been repeating a very similar liturgy for several decades.

The third and fourth types of evangelists, both already well represented among the patriarchs, are academics and entrepreneurs. Many of the books, articles, and academic conferences since the 1980s have predicted and continue to predict the advent of a digital society "within a few decades." Nicholas Negroponte and Alvin Toffler have often been seen as the two academic champions of the revolutionary prophecy, but there are sociologists, anthropologists, economists, psychologists, political scientists, and even historians who firmly believe that we're living in an era of complete transformation for humankind. The tone used in their writing (particularly in their titles and marketing blurbs) is often emphatic: "In the coming decades—the process has already begun—the entirety of our cultural inheritance will be transformed and re-edited in digital forms"; "we are living through the first stages of a world revolution as profound, in my view, as the invention of agriculture" (renewing the genre of excellent parallels discussed in Chapter 2), "the digital revolution is radically changing the way we make sense of our lives." For those who think that academics are irrelevant in constructing grand social narratives—a suspicion that often affects academics themselves—it is worth remembering that in this case, they performed the fundamental task of providing a chrism of scientific veracity to the digital revolution. This has benefited the same academics who, by becoming evangelists of the revolution, have founded consulting firms, exported the good news to various countries around the world through well-paid conferences, and gained popularity in the media. In doing so, their opinions have left the narrow confines of universities and scientific conferences and have been discussed by much wider audiences. In short, thanks to academics, the digital revolution gains scientific status and academics themselves, talking about the revolution, sell books and are interviewed in newspapers and on television. It is a sort of vicious (or virtuous, depending on who observes it) circle that is strangely driven by paper books, one of the products that the revolution should have swept away and that, instead, continues to represent the main instrument through which the results of academic research are conveyed. Negroponte, when asked why he wrote books but preached their disappearance, replied that he "wanted to permit all of us to imagine the new world," which he could not have done by expressing himself in other forms. In short, ordinary mortals need to be evangelized in a comprehensible way.

And books have been written by virtually all of the revolution's best-known entrepreneurs, the last and probably the most important type of prophets, messiahs, evangelists, and gurus to be examined. In fact, every country in the world has a pantheon of entrepreneur-gurus who have theorized the break between an analog and a digital world and have crowned themselves the protagonists of this transition. Many have lived in or passed through Silicon Valley, have absorbed its spirit, and have attempted to replicate it in their own countries. They have often been "stunned" by the digital revolution, which revealed itself to them suddenly but clearly and unequivocally, changing their way of thinking and living from that moment on. The best-known entrepreneurs are treated like stars in their respective countries and often abroad. They are welcomed as heads of state on their trips, recognized by the common people, they are ranked among the richest men (less frequently women, as will be seen) in the world. Many are inspired by the myths of Bill Gates and Steve Jobs, often aping their behaviors and aphoristic tendencies.

Some of these entrepreneurs are closely tied with politics. French multi-entrepreneur Gilles Babinet, as his Wikipedia page defines him, represents his country in the European Union's Digital Champions Group (a working group that promotes digitization) and writes books in which he theorizes the advent of a new era of human history due to the digital revolution. In the first decade of the 2000s, Italian "visionary" and

"BILL GATES, STEVE JOBS, ELON MUSK...HE COMBINES THE WORST OF EACH OF THEM."

Figure 11 The prototype of a digital guru.
© CartoonStock.com. Cartoonist Sidney Harris, uploaded on October 23, 2017.

"guru" Gianroberto Casaleggio predicted that by 2054 a form of direct democracy would be established globally and even founded a new digital political party, the Five Star Movement, which garnered over 30% of the vote in Italy's 2018 general elections. One could then cite dozens of new Chinese entrepreneurs, such as Jack Ma of Alibaba or Ma Huateng of Tencent—often exalted on the covers of major Western weeklies—whose prominent roles, however, have been downsized by the Chinese Communist Party in recent years.

Perhaps the most stimulating example of entrepreneur-evangelist, however, is the American (of Austrian parentage) Raymond Kurzweil. He works at Google and is considered the guru of transhumanism and the technological singularity previously mentioned. This is an interesting case because he can literally be counted among the prophets and messiahs since he proposes and promises to defeat human death by creating digital clones that can potentially live forever. The digital afterlife might be the most significant promise of the digital revolution and transformation or at least the one in which the quasi-religious character emerges most strongly, indeed aiming to replace traditional religions.

Among the names mentioned in this chapter so far, you will have noticed a dearth of female figures. Narratives extolling matriarchs and female evangelists of the digital revolution are missing or scarce, mirroring the generally strong gender disparity and lack of attention to diversity in the last of the contemporary revolutions. A 2021 UNESCO report, indicatively titled *To Be Smart, the Digital Revolution Will Have to be Inclusive*, highlighted the digital revolution's gender gap with a striking image: the illustration on the cover of the report depicts a woman attempting to board the "digital revolution" train, but she is forced to take a longer and more strenuous step than the men who are already comfortably on board. Thanks to an impressive amount of data, this report highlights how, in almost every country in the world, women have often been put on the margins of the digital revolution and have acquired fewer skills relevant in the digital market, how large digital corporations employ men in key roles, and how female entrepreneurs have less funding available to launch their own digital startups. And a similar argument can be made for digital evangelists of color: although over the years the prototype of US digerati has inspired various entrepreneurs in Africa and South America, the number of evangelists of color in the digital revolution remains limited to a small elite and a few isolated cases. The evangelists of the digital revolution have long been men, white, and American, only in recent years joined by some figures from Asia, but often inspired by the myth of the American evangelist. Women and people of color have not yet found their place in the digital kingdom of heaven; or rather, that place has been closed to them by a largely white, macho "culture" of revolution.

Having tried to define the categories and some of the more interesting characters, the question remains how the principal prophets,

Figure 12 Gender inequalities in the digital revolution. How to mind the gap?

© UNESCO. Cover of Bello et al., 2021, available at https://unesdoc.unesco.org/ark:/48223/pf0000375429

messiahs, evangelists, and gurus of the digital revolution come to be recognized as such. One form of investiture is through magazines. *Time* has bestowed the Man of the Year (Person of the Year since 1999) award to individuals who have distinguished themselves during the calendar year and, on several occasions, the award has gone to digital entrepreneurs: in 1997 to Andrew Grove (president of Intel), in 1999 to Jeffrey P. Bezos (Amazon), in 2010 to Mark Zuckerberg (Facebook), and in 2021 to Elon Musk (Tesla). *The Financial Times* gives out a similar award, Person of the Year, and many digital entrepreneurs have also made the grade in recent decades: in 1983 John Opel (IBM), in 1994 Bill Gates (Microsoft), in 1996 Rupert Murdoch (News Corporation), in 2005 Sergey Brin and Larry Page (Google), in 2010 Steve Jobs (Apple), in 2013 Jack Ma (Alibaba), in 2014 Tim Cook (Apple), in 2019 Satya Nadella (Microsoft), and in 2021 once again Elon Musk (Tesla).

Starting in 1994 and continuing for a few years, a specialized magazine closely linked to Silicon Valley, *Upside*, compiled rankings of the gurus of the digital revolution. In addition to the names previously mentioned, there were people not as well known to the general public (in every religion there are hierarchies of holiness). These included Netscape's Marc Andreessen, Masayoshi Son of SoftBank, John Chambers of Cisco, and Steve Case of American Online, but also the inventor of the World Wide Web, Tim Berners-Lee. Of course, many other magazines could be cited, such as *Forbes*, which has been compiling a list of the world's millionaires since 1987, a ranking literally dominated by digital entrepreneurs in recent decades. But I think these examples suffice to make the point: entrepreneur-evangelists are among the wealthiest and most famous public figures that contemporary society considers relevant.

Of course an evangelist is recognized by his tireless proselytizing. The digital revolution, then, is a form of global evangelization, and its gurus travel to different countries to spread the word: Steve Jobs—in addition to a trip to India in his late teens that strongly influenced his perspective and spirituality and which he never stopped talking about—used to travel at a frenetic pace. He travelled and walked so much that, in addition to his famous "uniform," a black turtleneck sweater and jeans, he had to wear comfortable shoes like his Birkenstocks sandals, one pair of which were sold in 2022 for nearly $220,000, since they retained "the imprint of Steve Jobs' feet, which had been shaped after years of use" (these sandals could literally fall into the category of relics, something that will be addressed later in the chapter).

Bill Gates has also been traveling to various countries around the world for years, predicting the classic "acceleration of the digital revolution" in each of them. Mark Zuckerberg has made several visits to China and, speaking in perfect Mandarin, tried—unsuccessfully—to convince the Communist Party leaders to open up to the digital revolution (or at least not to block access to Facebook in the country). These trips, and all of the entrepreneur-evangelists' activism, are not just dictated by a blind faith in the revolution and its "effects." There are personal reasons and advantages, products to sell, services to promote, and converging interests between politics, journalism, academia, and businessmen. Revolutionaries see themselves as "agents of history and historical necessity," as Hannah Arendt said. The various prophets, messiahs, evangelists, and gurus of the digital revolution feel like agents of history; they are described or describe themselves as interpreters and embodiments of the revolutionary spirit; they are men (and more rarely women) who put themselves at the service of an irresistible idea or ideology.

4.3 Heretics and Infidels

Does the digital revolution have enemies or, at least, opponents? In 2000, the Dutch philosopher and political scientist Evert van der Zweerde wrote the following:

> If the notion of "revolution" presupposes a powerful resistance that has to be overcome, an "ancien régime" of some kind, then we are certainly not witnessing a revolution. But if "revolution" means a rapid, radical,

irreversible, and shock-like process of change, I tend to hold that we are indeed witnessing a revolution, albeit a very smooth one. The reason for this smoothness, I think, is that, contrary to other revolutions, there are no major barriers or resistances, nor is there any political conflict related to it. No one opposes it really. Such reactionary and anti-modern forces like the Russian Orthodox Church are among the first to use the latest means of communication.

If it's true that contrary to other more conflictual revolutions, the digital revolution has often succeeded in bringing together all of the actors involved but over time two main categories of opponents or counter-revolutionaries have emerged: the heretics and the infidels. Heretics promote or follow a dogma that contradicts the "official" one, but continue to profess the religion; while infidels or unbelievers question the very existence of the dogma, in this case the revolutionary nature of the digital revolution. If there have been, and there are more and more, some members of the revolutionary faithful who can be called heretics, the infidels are indeed quite few in number.

Prime examples of heretics are, once again, academics. Even in the mid-1970s, some critical communication theorists highlighted the contradictions and problems of the digital revolution. They didn't question the fact that it is (or was) a revolution but rather, by criticizing it, they gave it further support or took its existence for granted. For example, Joseph Weizenbaum, a professor of computer science at MIT and a pioneer in the field of artificial intelligence, proudly claimed the title of "heretic" given to him by colleagues when he wrote one of the first books on the potential harmful effects of computers on human interactions. Vincent Mosco and Andrew Herman in 1980 and Robert McChesney in 2007 wrote (27 years apart) about a revolution that "we are witnessing" that would have negative consequences on many levels. Other authors who, especially in the last two decades, have initiated a critical turn towards digitization also believe in the existence and persistence of the revolution and of change. For Evgeny Morozov (defined on the back cover of one of his books as a "digital heretic"), Christian Fuchs,

Jack Qiu, David Lyon, Shoshana Zuboff, and many others, the digital revolution has in fact taken a wrong turn: it has fostered a system of political repression counter to the promises of societal democratization; it has promoted the formation of real capitalist oligopolies; it has exacerbated inequalities between those who have easy access to computers and the Internet and those who do not (the so-called digital divide); it has created insecure and underpaid jobs and, more generally, forms of exploitation of workers; it has led to an excessive consumption of energy resources in defiance of the green rhetoric of the same revolution; and it has endangered individual privacy while instead promoting a "culture" and a form of capitalism based on the surveillance and tracking of users. In short, digitization is chock full of broken promises and nightmares that are turning into reality, but it is still a revolution and a rupture that is having tangible, albeit dystopian, effects on contemporary societies.

Surprisingly, some heretics can be found among evangelists or former evangelists. For example, negative comments on the consequences of digital technology have even appeared in the pages of *Wired*, especially in recent years but also in a 1998 article significantly titled "Is the Revolution Over?" This article highlighted how, at a number of Silicon Valley companies at the end of the millennium, the revolution had been translated into a way to exploit workers, who began to distrust revolutionary language, considering it "just a management ploy to squeeze more work hours out of employees"; furthermore, the attractiveness of the ideology of revolution seemed to be declining, and the same "stinking revolutionaries," obsessed with producing something innovative, were defined as pathetic. Also in 1998, another landmark magazine of the digital revolution, *Upside*, published a highly critical article by a former evangelist, journalist Michael S. Malone. In addition to masterfully summarizing many of the stereotypes and utopian discourses related to the revolution (from its undoubted existence, to its ability to subvert the established order, to its transcendent dimension), Malone spoke of a "technofascism" that had accompanied the digital revolution up to that point: no-one *can* deny it or even oppose it, and those who do so are accused of being backward, of not understanding the future, of being an antichrist figure. It is also interesting that Malone included himself among the techno-utopians and fascists and, in the years following this article, he continued to argue for the existence of the digital revolution and to embrace its ideology. The heretics of the revolution are self-confessed fascists; they are often aware of adopting propagandistic language and tending to exaggerate the facts, but after all they're doing it for a good cause: to implement the revolutionary utopia.

Even Mr. and Mrs. Toffler seemed to backpedal during the dramatic bursting of the dot.com bubble, admitting that they had exaggerated their optimistic predictions, which needed to be downsized starting with the fact that the digital revolution was "not the only source of fundamental change." In 2013, even Bill Gates, who in recent years has increasingly focused on his humanitarian foundation, while still professing to be "a huge believer in the digital revolution" and its value of connectivity for primary healthcare centers and schools, gave a negative assessment of Google's plan to use high-altitude balloons to bring Internet connectivity to Africa: "When you're dying of malaria, I suppose you'll look up and see that balloon, and I'm not sure how it'll help you. When a kid gets diarrhea, no, there's no website that relieves that." In 2015, another VIP of digitization, Google executive Vinton Cerf—often presented as one of the fathers of the Internet (and therefore a patriarch of the revolution)—invited people to print their photos, emails, and their most precious documents to avoid losing them. If you think about it, it's a complete about-face, one in which an analog solution is proposed to remedy the flaws of the digital revolution: in this case the constant risk of losing one's data. In short, a transition from bits to atoms.

Recently, it is exactly this type of "return" or "revenge" of the analog that has become one of the most discussed counter-revolutionary trends, one that has involved consumers and that, naturally, has been welcomed by some companies: for example, in the music industry, where the vinyl record and tape cassette have seen a major comeback, while the field of photography has seen increased sales of very popular analog camera models, such as Polaroid or Lomo. To what do we owe this return of analog, which, for the acolytes of the digital revolution, may seem like a setback or simply the return to a past that was believed to be outdated and that the revolution should have swept away? These are phenomena that have to do with fashion, with the trend toward vintage, with the growing social relevance of nostalgia, but that complicate the "classic" narrative of the digital revolution as a clean break with the previous world (the revolution's most significant mantra). In today's increasingly complex media landscape, seemingly abandoned analog tools and new digital media can in fact coexist and even promote the revival of practices, devices and economic models from the past. Among the reasons for the return of vinyl records and analog cameras we can recognize a sort of "reaction" against the digitization and dematerialization of formats, and in favor of using traditional physical formats instead. This is why the "return" of the analog can be numbered among the heretical discourses, but not one that is unfaithful to the digital revolution: in fact, it doesn't question the existence of the revolution but rather identifies niches of counter-revolutionary resistance that are successful precisely because of this opposition. In short, without the digital revolution there wouldn't be any nostalgia for analog.

Similar heresies can be traced in recent forms of resistance to digitization: the so-called digital disconnection and the digital detox. A small minority of people, primarily in rich countries, from time to time decide to take a break from digital communications, to unplug and do a "digital detox": it is an attempt to control their presumptive dependence on digital technologies and to reappropriate their own lives and time. But, in most cases, these "sinners" allow for an inevitable return to their "normal" digital consumption after a period of rest and so it's just a form of temporary heresy.

In contrast, the infidels have put forward a thesis that can be described, not without irony, as disruptive: the digital revolution *is not revolutionary*. First, it isn't revolutionary because it doesn't involve a radical change in the habits of human beings, unlike other revolutions of the past. Robert Gordon, the economist who is perhaps best known for his denialist positions on the digital revolution, has argued that the century between 1870 and 1970 was far more significant in shaping people's daily lives: suffice it to say that, in this time frame, major transportation, electricity, gas, telephone, water, and sewage networks were in fact built or consolidated. He goes on to point out that these have had a far greater impact in transforming people's lives and habits than something like computer networks, the Internet, or mobile phones.

Another argument put forward by the infidels claims that the digital revolution cannot be considered revolutionary because it doesn't completely supplant the old world. As Negroponte put it, it doesn't eliminate atoms but rather has led to a proliferation of tangible objects and infrastructures; it doesn't constitute a new economic paradigm but only represents an acceleration of old constructs and trends, as the formulas of neo- or turbo-capitalism show; it has not replaced previous forms of communication, as various prophecies had claimed that it would. These positions have often called for looking at digital revolution and then digital transformation as evolutionary rather than disruptive processes, traversed by gradual, slow, and long-lasting developments.

But the infidels and deniers are few. More significant are the heretics who, after all, share a common ideology with patriarchs and prophets: they all firmly believe in the *existence* of the digital revolution and in its transformative nature for humanity in various fields. They are certain that it is about to happen or has already happened, and they popularize the dogma of its existence through books, conferences, interviews, keynote speeches, and documentaries. The only difference is that in one case they exalt the digital revolution, while in the other they embrace apocalyptic visions regarding its present and future. But countering the positive side of the revolution can reinforces the revolution itself: according to Koselleck, once legitimized, every revolution needed to "continually reproduce its foe as a means through which it could remain permanent." It is useful for digital revolutionaries to be able to counter the alternative and pessimistic dogmas of heretics because by surrounding themselves with real or potential enemies, the evangelists of the revolution can reaffirm their function and usefulness, they can show that progress still needs to be made in order to impose an ideology.

4.4 Relics

Various religions venerate objects, bodies or remains of famous people that are often called relics. And so, even the digital revolution has its relics. I'm not just talking about cult movies, the first prototypes of computers or mobile phones preserved in museums, or even the statues that the cities of Belgrade, Bucharest, Budapest, Istanbul, Odessa, and probably several other cities have decided to dedicate to Steve Jobs or the previously mentioned monuments erected in honor of Alan Turing. The relics I'm talking about are more ordinary and don't belong to deities, saints or exceptional people, but to all of us: they are the dozens of digital devices we use and have used throughout our lives.

Although a list of the most significant digital technologies risks being partial, there are some objects that have symbolized the revolution more than others: in the 1940s and 1950s these were the first computers, such as ENIAC and UNIVAC, but also the transistor, to which we will return later. In the 1960s, digital watches invaded the world market, becoming perhaps the first massively popular product of the digital revolution. In the 1970s microelectronics exploded, and along with them some fundamental innovations, such as automation, miniaturization of products, and telematics itself. In the 1980s, music CDs and the first personal computers marked the entrance of the digital revolution into the home. That was also the decade that marked the silent gaming revolution: think of "mythic" consoles such as the VIC-20 or Commodore-64 (I am sure many readers will feel slighted because I have forgotten dozens more). Then in the 1990s, there was the PlayStation, which became a mass phenomenon in the new millennium, so much so that the documentary *From Bedrooms to Billions: The Playstation Revolution* was made about it in 2020. Also in the 1990s, some new applications, such as virtual reality and augmented reality technologies, seemed imminent (and yet they still haven't managed to break through), but above all it was the decade in which the Internet became popular thanks to the World Wide Web and the first generation of mobile telephones exploded in Western countries. The first 10 years of the 2000s saw the vertiginous growth of devices for listening to music, taking photographs, making calls, and working in a way that was increasingly on the go and wireless—wireless connections were the real godsend of the revolution. The second decade of the 2000s saw the rise of cloud computing in the field of content storage, artificial intelligence applied to various aspects of daily life, and the wearable devices that track our daily movements. And who knows how many new relics the digital revolution is poised to churn out in the future: from 5G applications to quantum computing (a new generation of much more powerful and high-performance processors that seem to be bringing Moore's Law back into vogue), from new foldable smartphones to the ever-present prediction of under-the-skin chips that has been continually and invariably repeated for several decades.

Curiously enough, among the objects revitalized by the revolution one finds old analog devices that have been revived through digital technology and, as always, given new names (the digital revolution not only continually changes its name but also rebaptizes the objects it consecrates). Currently the adjective *smart* or, a few years ago, the suffix *e-* represent two of the many magic words obsessively used to revitalize old technologies and incorporate them into the revolution: household appliances, cars, cities, watches, televisions, and telephones become smart; just by adding an *e*, mail, commerce, government, and books are digitized—constituting new examples of progress and innovation.

Even more interesting are some items that were supposed to break through and take part in building the imminent future of the digital revolution but that actually turned out to be resounding failures. Interactive digital television, optical discs, high-definition as well as 3D digital television, electronic diaries, the attempt to save audiocassettes or VCRs by putting "digital" versions on the market, pagers (defined in 1996, the "revelation of the information revolution"), Steve Jobs' Next workstation, notebooks, DAB as the new digital standard for listening to the radio, and dozens of other objects that crowd consumer's attics, company warehouses, or even landfills around the world. These little-used products or those that quickly disappeared from the market are the clearest evidence that the digital revolution can fail, but strangely (or strategically) they often become cult objects among enthusiasts, collectors, and nerds who compete to snatch them up or discuss their "mythic" performance on the Web in an inevitably nostalgic tone.

Of course, not all digital devices are or will be relics. In this sanctification process, manufacturers play an important role because they push their devices through advertising or (re)creating stories around them, establishing a hierarchy between more or less symbolic objects. But it is then the consumers who, by doing things like camping out in front of stores the night before the launch of a new smartphone model, attribute to a digital relic a value relevant to their existence, exhibit it in front of others, and incorporate it into their lives. Another powerful player in the device-recognition process is the media. Perhaps one of the most interesting examples is something already mentioned: *Time* magazine's awarding of its 1982 Man of the Year prize to the computer because, "above all, it is the end result of a technological revolution that has been in the making for four decades and is now, quite literally, hitting home"—of course, hitting home is an apt metaphor because those were the years when personal computers were starting to enter millions of US homes. According to *Time*:

> It would have been possible to single out as *Man of the Year* one of the engineers or entrepreneurs who masterminded this technological revolution, but no one person has clearly dominated those turbulent events. More important, such a selection would obscure the main point. TIME'S *Man of the Year* for 1982, the greatest influence for good or evil, is not a man at all. It is a machine: the computer.

In this passage there is an evident transcendent and superhuman dimension: the digital revolution, in addition to being promoted and conveyed by human actors—such as patriarchs, evangelists and heretics—is also driven by non-human actors, like digital devices. Moreover, these objects have agency: that is to say, they have a life of their own, they influence and stimulate the revolution or simply embody it in the most natural and evident way for the acolyte-consumers.

Another much-cited example is Steve Jobs' presentation of the iPhone in 2007 when the CEO of Apple, before an audience of the faithful waiting to be amazed, laugh, and swoon, masterfully condensed the essential concept of the digital revolution. In the first three and a half minutes of the presentation (easily found on the Web), Jobs traced a brief history of the digital revolution, which had proceeded thus far by disruptive innovations and of which Apple had been the main protagonist: in 1984, with the Mac, it revolutionized the computer industry ("it didn't just change Apple, it changed the whole computer industry," Jobs argued); in 2001, with the iPod, it revolutionized the music industry ("it didn't just change the way we all listen to music, it changed the entire music industry"); and in 2007, with the iPhone, it revolutionized the telephone industry ("today Apple is going to reinvent the phone"). The perfect storm or the perfect narrative: the evangelist, now almost a patron saint, presenting what would become one of the most famous relics of the digital revolution and, in doing so, guiding the faithful in the interpretation of what is happening with some key words (in addition to revolution, another term used obsessively is breakthrough). It is no coincidence that this presentation was called "the second coming of Christ" and that the iPhone began to be called the "Jesus phone."

"I could re-design that for you."

Figure 13 Steve Jobs in heaven, a God among Gods. Part 1.
© CartoonStock.com, cartoonist Jim Bertram, uploaded March 6, 2014

These relic-devices allow the user to enter into communion with, if I can say so, the digital revolution. They are blessed hosts thanks to which the revolution reinforces the faith of its acolytes, the tool that consumers have to get closer to the Pan-like God of the revolution. Thanks to these devices, you can in fact "join the digital revolution." This mantra is obsessively repeated for almost every device, but a fun case in point is the digital watch. A quick search in the archives of *The New York Times* shows that between 1976 and 2015 as many as four advertisements for different brands of digital watches used the same slogan: "join the digital revolution." The digital watch then becomes

Figure 14 Steve Jobs in heaven, a God among Gods. Part 2.
© CartoonStock.com, cartoonist Chris "ROY" Taylor, uploaded March 6, 2014

an access portal to the revolution's heavenly kingdom, a narrative that was repeated, substantially unchanged, for 40 years. The mantra about joining the digital revolution is not the only one that has accompanied the presentation of these relic-products for decades. In fact, there are

"Sorry to call you up early Steve, but our network just crashed."

Figure 15 Steve Jobs in heaven, a God among Gods. Part 3.
© CartoonStock.com, cartoonist Kelly Kincaid, uploaded October 10, 2011

others that not only repeat themselves over time, but have also been attached to different products. For example, each new digital device or service is regularly referred to as "the biggest" or "most advanced innovation of the digital revolution," the one that "started the digital revolution" or is "driving it." Or components largely unknown to the general public, such as the transistor, are often extolled retrospectively and their invention hailed as a "miracle" that generates a "mix of wonder and envy." And of course there are illustrious parallels with the past: a number of revolutionary stereotypes are condensed in a single Fujitsu advertisement from the 1980s, which aimed to promote their new ISDN services for fast Internet. The primary payoff line was "Fujitsu is pleased to announce the end of the world . . . as we know it." ISDN, a standard that would quickly be superseded by ADSL, was touted as:

> the logical conclusion of the 19th century. We've got a million new things that century never had, of course—like cars, planes, radios, TVs, satellites and personal computers. But we've still got the same old minds. Those minds were formed, by and large, by the Industrial Revolution. That Revolution took us off the land, put us in the city, gave us jobs, money, education, progress and lots of problems. And that's where we are today. But that's not where we're going to stay. Because ISDN has arrived. ISDN? ISDN is the rising force in the next revolution—the Information Revolution. It's a complex range of integrated computer and communications technologies, whose effect will be very simple: It will put an end to the world as we know it.

Let's play a game, similar to the one I proposed that you play in the Introduction: replace the term ISDN with any other random digital technology from the last few decades and you will get some advertising slogan, a CEO-evangelist's statement, the words used by a politician at a conference or in a parliamentary chamber. Although digital devices change over time, they are often presented in a similar way, even using the exact same words. The arguments that are supposed to certify their revolutionary pedigree are repeated, tediously, through the decades. But the digital revolution needs artifacts, relic-devices to materially anchor all the narratives it has constructed: in this case, the objects sold and bought in the billions are *evidence* of the unshakable faith of companies and consumers, but also a physical sign that the digital revolution will continue in the future (one of the meanings of relic is, after all, precisely that of an object that has survived from the past). This faith is unshakable, and even in the face of the rapid replacement of contemporary technologies, which renders a newly purchased item obsolete, we must believe in the next digital device—the one that will, invariably, allow us to join the revolution.

4.5 Meccas and Shrines

In addition to humans and objects, the digital revolution has produced
other entities to be remembered or revered: these are physical places
where the revolution began, where its protagonists studied and worked,
and where its devices are produced. Although this spatial and physical
dimension is paradoxical for a revolution that promised and promises to
be virtual, and thus to be in all places and nowhere specifically, in reality
the revolution's meccas and shrines are essential from a quasi-religious
perspective. They are symbolic places but at the same time geograph-
ically located and therefore make the revolution tangible. They are
places that can be replicated, which contribute to the work of spreading the good word of the revolution because they must be visited and
in fact are the destination of real "pilgrimages."

It seems almost a given to cite Silicon Valley as one of the shrines,
or rather *the* shrine, of the digital revolution. Over the years, patriarchs, evangelists and heretics have converged on Silicon Valley. Many
of the relics that have entered the pockets, homes, and offices of billions of people were invented there, and it's where some of the most
significant companies of the digital revolution, represented by the
acronym GAFAM (Google, Amazon, Facebook, Apple, and Microsoft),
have their headquarters. Silicon Valley has forged the so-called Californian ideology that underpins many of the narratives built around the
digital revolution. Beginning in the 1960s, previously distinct ideologies
such as counter-cultural movements, anti-statist and liberalist political claims (which brought together America's new right and new left),
the cult of individualism, some neoliberal economic doctrines, and a
good dose of technological determinism and utopianism intersected in
this relatively circumscribed region of California. These are all ingre-
dients that are still present in discussions about digitization and digital
transformation today.

However, Silicon Valley hasn't always been a digital mecca. In the
1960s, this region was an area of high immigration where intensive

agriculture was practiced. In the 1970s, several microelectronics and semiconductor companies settled there, enough of them that the first definition of "Silicon Valley" was coined in 1971 in the industrial tabloid *Microelectronics News* to highlight the dominance of three major companies in the industry: Fairchild Semiconductor, Intel, and National Semiconductor. In the 1980s, a number of biopharmaceutical companies moved or established their headquarters in the Valley. Silicon Valley only started to take on its modern form in the 1990s, following the emergence and proliferation of various digital companies that then, beginning in the early 2000s, gained relevance and dominance in the global marketplace.

But Silicon Valley isn't the only and perhaps not even the first place on the planet devoted to the digital revolution. For example, in the 1980s and 1990s, Japan's Kyushu was nicknamed "Silicon Island" because it was home to about a quarter of the country's entire semiconductor production and to the headquarters of some of the most symbolic companies of Japan's digital revolution, including Sony and Mitsubishi. Over the years, Kyushu's relevance has declined and new silicon valleys have emerged, for example Fukuoka. A few years later, the same fate befell the Chinese district of Zhongguancun, where some of Beijing's major universities are now located. In fact, this area was established with the explicit intent of creating the Chinese Silicon Valley and inherited a series of cultural influences, such as neoliberal logic, the idolatry of innovation, and flexible—one might even say unstable—working conditions. However, Zhongguancun hasn't been able to replicate the success of its American counterpart, and today the place that is considered China's Silicon Valley is located in Shenzhen (the city that is home to global companies such as Huawei, Tencent, TP-LINK, and many others) where, it is often repeated, the next digital revolution will start or has already started.

Another high-growth area is in southern India, in Bangalore. Again, this city is referred to as "India's Silicon Valley" or the "IT capital of India," and like the cases presented above, a fair number of the country's digital companies, such as ISRO, Infosys, Wipro, and HAL, are headquartered there.

Reproducing Silicon Valley is obviously a global cliché and all you need to do is search on "The Silicon Valley of . . .," adding any country to realize that. There is even a kind of competition to be appointed (or self-appointed) the new Silicon Valley: for example, Dublin, Stockholm, Barcelona, and Berlin, among others, are competing for the scepter of Europe's Silicon Valley.

In addition to the influx of public and private investment and the possibility of attracting the best talent, Silicon Valleys generate quasi-religious tourism, just like other cities symbolic of the major religions (think of Rome, Jerusalem, Mecca, the holy cities of Hinduism, and many other examples). Heads of state and politicians from various countries visit them in order to understand the digital future and be inspired by it, entrepreneurs of all sorts spend periods in these areas to absorb the culture and teachings and then apply them to their own sectors; however, increasingly it is ordinary people, believers and acolytes of digitization who organize *sui generis* pilgrimages to the headquarters of Google, Amazon, Tencent, Samsung, or the visionary company of the moment.

Yet despite the proliferation of Silicon Valleys around the world, to date no other place has the charm and appeal of the original Silicon Valley, the one located in California. For example, a 2006 article by one of the best-known reporters of the US edition of *China Daily* posed the question: "But what does the digital revolution, which has already brought dramatic changes to our lives and work, have in store for us? I could have found some answers if I'd had the chance to visit the Home of the Future at the main headquarters of Microsoft Corp." Predicting the future of the revolution is easy, just go directly to the oracle on duty. Perhaps even more paradoxical is the biography of a not-so-well-known digital entrepreneur on the back cover of his *Manifesto of the Digital Revolution*:

> Jawad Essadki is a serial entrepreneur and visionary who has been living in the Silicon Valley for the last decade, observing the digital revolution as a consumer. In 2015, his observations and studies led him to a "common mortal" perspective and he chose to share his view of the industry with the rest of the world . . . with the goal of giving all of us some perspective on the future.

A number of the digital revolution's stereotypes and quasi-religious keywords are on the verge of caricature: the visionary entrepreneur who can now show humanity the way of the future because he immersed himself in the sacred river of Silicon Valley, where he was able to closely observe the revolutionary truths after being unfailingly thunderstruck on the road to Damascus.

In addition to Silicon Valley and the headquarters of large companies, research centers are also symbolic locations where the revolution happens: to cite just two examples, some of the inventions crucial to digitization came out of the previously mentioned Bell Labs and in an underground corridor of Geneva's CERN—one of the world's most famous physics laboratories, which has revolutionized human knowledge through its discoveries—you might come across a plaque that reads "Where the Web was born." The digital revolution also happens in unexpected, secondary places, but it is or should be always celebrated publicly.

There are other specific places to officiate the ritual of the revolution: the high-tech trade shows (among the best known are Comdex, Computex, and CeBit), the hundreds of dedicated public conferences, the galas like the Computerworld Smithsonian Awards in Washington, where the "true heroes of the information revolution" are honored. The most famous participant in these meetings was undoubtedly Bill Gates. Despite not being a great speaker, all Gates had to do was get on stage and the faithful attendees were hanging on his every word, trying to grasp the future direction of the revolution. Inevitably, the media described his speech in quasi-religious tones as, "the Messiah's return to Mecca."

After being conceived in Silicon Valley or research centers and launched in style at high-tech trade shows, the relics of the digital revolution are displayed in specialty stores for purchase. And the stores of the major digital companies are veritable shrines to revolutionary ideology, placed in strategic locations in major cities, they are large transparent buildings that allow acolytes to observe and desire the latest

wonder; even heretics are allowed to take a look inside and be convert-
ed. The most striking reflection on the subject probably comes from
The Simpsons, the well-known cartoon created by Matt Groening, in an
episode that was first aired in the United States in 2008. The whole family visits a Mapple store (evidently a parody of Apple stores), where
suddenly a message from the founder Steve Mobs is announced: the
acolytes inside the store go into a frenzy and exclaim "He's a genius!"
"He's like a God who knows what we want!" But, as usual, Bart invents a
ruse to intercept and distort the message that should have "completely
change[d] the way you look at everything," thus unleashing the wrath
of consumers and store employees. Through this caricature, these initial minutes of the episode manage to give an account of the sacred
dimension of the Apple store, where products and services are sold at
astronomical prices (this is what Bart tells the acolytes and what Lisa
intimately experiences until she gets a meeting with Mobs-Jobs), but
where there is also a daily ritual: the display of relics.

Finally, there are other interesting locations, this time dedicated to
the preservation and memory of the digital revolution: museums and
non-profit organizations. For example, the Computer History Museum
in Mountain View, California has organized and continues to organize exhibitions about the digital revolution and transformation, such
as the notable 2011 exhibition *Revolution: The First 2000 Years of Computing*, which aimed to tell "the history of the computing revolution and
its ongoing impact on society worldwide." The mission of the Charles
Babbage Institute, founded in the late 1970s, is to study "the historical
roots of the information revolution" and thus prepare for its future
developments. More generally, museums dedicated to computers or
other digital technologies are flourishing, and in national science and
technology museums, the space and focus on digitization is growing steadily. In many of these museums the "first" computers and
the "first" mobile phones are preserved, or the exceptional (and very
similar) biographies of the heroes, engineers, and entrepreneurs who
"first" understood the developments of the revolution are recounted.

Preserving and promoting the memory of the revolution is not an inconsequential act, it has the effect of placing the digital revolution on an "official" and recognized historical trajectory.

4.6 (At Least) Four Commandments

Between the lines of this chapter, at least four commandments of the digital revolution have emerged (every religion, after all, has its own rules to follow), but readers will probably have spotted many more.

First: the digital revolution was, is, and will be. With the exception of a very few "infidels," in the public discourse, the digital revolution is a revolution that has happened and a transformation that is underway and will hold many surprises in the future. Its existence is not questioned even by those who are today its harshest critics (called "heretics") who would like to change its direction, who point out its weaknesses, but who do not question it. Faith in the existence and continuation of the digital revolution is therefore blind, unquestionable, and dogmatic.

Second: the digital revolution is multi-faith and tolerant. The revolution has borrowed metaphors and key terms from various religions, transforming and adapting them to its own needs: prophets, messiahs, infidels, relics, and holy cities are all present in the major world religions. In the universe of the digital revolution, Christianity, Islam, Hinduism, Judaism, and many other faiths coexist peacefully, contrary to what happens in the "real" world. One of the main reasons for this form of tolerance and religious coexistence lies in the fact that the digital revolution is global by nature, it wants to penetrate all countries and cultures without resistance and, to do so, it must also adapt to different religious contexts. The religious terms and metaphors that have served as titles of this chapter's paragraphs are not just a *divertissement* for the author but are intended to account for the revolution's flexibility, mimetic capacity, and adaptability.

Third: the digital revolution is an easy religion to profess. Patriarchs and patron saints have marked the way forward, acolytes are guided by the words and tireless evangelization of prophets, messiahs, evangelists, and gurus; it is enough to buy certain items or visit certain places to incorporate the revolution automatically, there are no particular obstacles or enemies to overcome—except perhaps not having enough money. Good acolyte-consumers have only to let themselves be meekly guided by such authoritative figures (and by the intensive convergence of interests) and behave well: "Don't be evil," Google tells us in one of its best-known slogans.

Fourth: the digital revolution is continually reborn. There is a profound tension in revolution: its need for constant renewal in the face of fixed patterns and clichés that are repeated over time. Unlike true religions that have had well-established doctrines and rituals for millennia, the religion of the digital revolution has a constant need to create new evangelists, new relics and meccas to worship, or to reinvent its tradition by rediscovering new patriarchs and patron saints. There are perhaps only two exceptions to this: *Wired*, which since the 1990s has remained the Bible for those who want to read the word of the revolution, and the figure of Steve Jobs who, even and especially after his death (the revolution has its martyrs, as we have also seen with McLuhan), has embodied the prototype of the revolutionary hero. However, this variability necessitates fixed structures: even decades later, some slogans or mantras persist, some stereotypical formulas are repurposed and passed from character to character or device to device, and the wonders as well as the concerns associated with the revolution are essentially the same. *Amen.*

Endnotes

Cited in relation to the quasi-religious form of the digital revolution are: Mosco, 2004; Natale and Pasulka, 2020, especially in the introduction; Winner, 1984 (quotations at pp. 585 and 593). Also relevant to the discourse developed in this chapter are Alexander, 1990 (with an excellent analysis of the religious language that some of the major US magazines have used to talk about the advent of the computer between the 1960s and the 1980s), and Tsuria, 2021. Finally, regarding religious faith in technologies the amount of literature is impressive: I have found inspiring, and with a historical slant and also some references to the digital revolution, Hinkson, 2013–2014; Noble, 1997; Supp-Montgomerie, 2021. See also the special issue of *History and Technology* entitled "Religion and Technology" edited and introduced by Karns Alexander (2020). On the semantics of "to believe," see *The Compact Edition of the Oxford English Dictionary*, Volume I (1976, p. 196). Matt James' blog is titled *How to Keep Faith in the Digital Revolution*, while the Reformed Church of Google, with all of its prayers and commandments, can be found on the website https://churchofgoogle.org/ and on Reddit. Kopimism has a Swedish-language website but also an introductory page in English: https://kopimistsamfundet.se/english.

In Section 4.1, the meanings of patriarch can be found in *The Compact Edition of the Oxford English Dictionary*, Volume I (1976, p. 557). The quotations regarding mathematician-patriarchs are from the following sources: the Doodles dedicated to Boole (November 2, 2015), Lovelace (December 10, 2012), and Shannon (April 30, 2016); Alan Turing was also honored with a Doodle on June 23, 2012, but the quotation used is found in Graham-Cumming, 2012. The quotes regarding entrepreneur-patriarchs come from a variety of sources: those of Andrew Grove in Garner, 2016 (there is also the hagiographic-toned bestseller about Grove written by Jackson, 1997); the one about Paul Allen was from "Paul Allen. Obituary," 2018; that on Douglas Engelbart in "Doug Engelbart," 2013. On "new patriarchs," see Little and Winch, 2021. On academic-patriarchs, in addition to the others previously cited, the reference is to Drucker, 1968; Machlup, 1962; Masuda, 1980; Porat, 1977. On the patriarch-hackers, we highlight the work of Levy, 2010, p. viii. On the hacker look, and more generally on the religious and hagiographic aspects of the history of the Internet, see Russell, 2017. There are dozens of articles about McLuhan in *Wired*, but three are of particular interest: Levinson, 1993; Wolf, 1996a, 1996b. Wolf also wrote perhaps the most interesting "biography" of the magazine (2003), where, on pages 70–71 he recalls McLuhan's role in the visual manifesto of *Wired*. McLuhan's investiture was also confirmed and commemorated in later articles, such as Shachtman, 2002. The text also quotes from McLuhan, 1964.

Prophets, evangelists, messiahs, and gurus dealt with in Section 4.2 belong to different religious traditions. Prophets are the most transversal figures because they appear in Judaism, Christianity, Islam, and other religions. The prophet speaks in public in front of listeners, instead of or in the name of someone (usually a God), and are able to predict the future (see *Oxford Learner's Dictionaries*). Messiahs are also present in the three major monotheist religions where they are presented as an expected liberator or savior of an oppressed people, country, or cause. In the Christian tradition the Four Evangelists are charged with spreading the gospel or "good news" and forming new ecclesiastical communities but, by extension, their words in the canonical gospels are also seen as infallible and they are considered trustworthy (see for example http://www.dictionary.com). Gurus are Hindu religious teachers and spiritual guides in matters of fundamental concern (See *Merriam Webster*). The definition of digerati is taken from Wikipedia page dedicated to the term and the article in which it first appeared (Markoff, 1992). For the first category of politicians, the North Korean leader Kim Jong-un's positive and proactive attitude about the digital revolution is confirmed by Lee, 2011. A reflection on the political need to legitimize investments in digital technology can already be found in Humphreys, 1986. For the category of journalists, the text refers to Walter Isaacson's bestseller: *Steve Jobs* (2011a) and *The Innovators* (2011b). For the academics, three examples of emphatic and peremptory statements about the digital revolution are cited: the first is from literary historian McGann, 2005, p. 72; the second from British anthropologist Hart, 2009, p. 24; the final one is from the back cover of the successful book by Weinberger, 2007. Negroponte's statement about books is found in a review by Rud, 1997, p. 31. Finally, for the category of entrepreneurs, reference is made to Babinet, 2014, 2016. For Gianroberto Casaleggio, see Casaleggio, 2004 and the short video from 2008 *Gaia. The Future of Politics*, available on YouTube. Available online and as a download is the Ray Kurzweil Reader (n.d.), a collection of essays and conference presentations on the digital revolution and redefining the concept of death. Under this respect, a key text is Sisto, 2020. Cited in reference to the digital revolution's gender gap is Bello et al., 2021. Regarding digerati of color, see Barber, 2006. The recipients of the person of the year awards from *Time* and the *Financial Times* are easily found online (and constitute an interesting source for understanding some contemporary "myths"), while it was necessary to do archival research to find the material from *Upside*. On the sale of Steve Jobs' sandals see Anguiano, 2022. The reference to Hannah Arendt at the end of the paragraph is found in her book *On Revolution* (1990, p. 53).

The passage from van der Zweerde that opens Section 4.3 on heretics and infidels is quoted in Lüthy, 2000, p. 66. For the different meanings of heretic/heresiarch and infidel, see http://www.dictionary.com and *Oxford Learner's Dictionaries*. The critical text on the role of computers written by Weizenbaum was published in 1976; on the fact that he was claiming the title of heretic, see Markoff, 2008. Among the critical theorists cited see Mosco and Herman, 1980, and McChesney, 2007. The title "digital heretic" is given to Morozov on the cover of his 2013 book. "Broken promises" is in fact part of the title of an article by Albrecht, 2014 which examines many of the topics cited. Regarding the heretics who are ex-evangelists, the dark outlook offered by *Wired* is found on pages 102 and 106 of Bronson, 1998; the confessional article by Malone, 1998; Alvin and Heidi Toffler express their partial retraction in Toffler and Toffler, 2001; Gates expressing skepticism regarding Google's project is found in "Gates Sceptical of Google's Balloon Internet Connectivity Helping the Poor", 2013; Cerf's recommendation to print photos and emails received ample international attention (Sample, 2015). On the return of the analog in the digital era, see Balbi and Magaudda, 2018. On the seeming paradox of the explosion of nostalgia for the analog, see Boym, 2001, p. xvi, who reminds us that "outbreaks of nostalgia often follow revolutions." On digital detox, see Syvertsen, 2020. Among the infidels, I cite the thesis of Gordon, 2016, but also other economists like Krugman (2023) made pessimist predictions about the transformative power of digitization and the Internet specifically. Finally, Reinhart Koselleck's reference to the need for a revolution to produce enemies is found in his 2004 book at p. 56.

In Section 4.4 on relics, the quotes about computers like *Time*'s Man of the Year are found in Friedrich, 1983, and those from Steve Jobs in his presentation of the iPhone are taken from one of the many videos of the event found on YouTube, which I recommend you watch. A scientific article central to understanding the quasi-religious dimension of the iPhone presentation, from which the quote regarding the second coming of Christ is drawn, is that of Campbell and La Pastina, 2010. The advertisements for watches in which the same slogan is used again 40 years later were published in the following editions of *The New York Times*: January 11, 1976 (the brand was Federated Electronics); September 16, 1998 (Bloomingdales); December 16, 2005 (here even used by several different brands sold at Macy's, including Diesel, Casio, Kenneth Cole, and Fossil), March 24, 2015 (Stauer). "The most dramatic," "The most advanced," and similar expressions are used in articles about the ENIAC computer (Farrington, 1996), fiber optics (Crisp, 1983), the Nokia 9000 Communicator (Keegan, 1996) and dozens of other objects. Expressions like "Triggering the start of the digital revolution" or "Driving the digital revolution" are used to talk about integrated circuits ("Intel Celebrates 40 Years of Digital Revolution" 2011), fiber optics (Dobbin, 1996), and many other products. We find transistors spoken of as a "miracle" in Elliott, 1997, and of "wonder and envy" celebrating the 1997 Man of the Year award to its inventor Andrew Grove (Isaacson, 1997, p. 48). The Fujitsu advertisement is taken from "Fujitsu. A Survey of Telecommunications" 1987.

There is ample scientific literature on the culture and history of Silicon Valley with which Section 4.5 opens. A "classic" is the work of

Barbrook and Cameron, 1996. The two books that are perhaps closest to
the spirit of this text, if only because they aim to trace the genesis of the
intellectual foundation of Silicon Valley companies, are by Daub 2020,
and Pugh O'Mara, 2019. On the other Silicon Valleys around the world
see Itoh, 2001 (for Japan), Negro and Wu, 2020 (for China), and Collato,
2010 (for Bangalore). On the necessity and benefits of visiting the headquarters of successful digital businesses see Li, 2006, and Essadki, 2015,
where I quote the back cover. Regarding the other places important
to the revolution, the citation regarding the Smithsonian gala is taken
from the awards speech given by the organizer Daniel Morrow in 1999
("Koch Honored as a True Hero" 1999), and on Bill Gates as the Messiah returning to Mecca see Schwarzenegger and Balbi, 2020. *The Simpsons*
quotations are from the seventh episode in the series' 20th season entitled "MyPods and Boomsticks" and was first broadcast on November 30,
2008. Among the places where the revolution is memorialized, the first
citation is found in Marsh, 2013, p. 640, and the second in Gullard, 1981,
p. 263.

Conclusion
Who Needs the Digital Revolution and Why Does it Keep Going?

Two fundamental questions remain to be addressed: who needs (and is served by) such a perfect, complete, irrefutable social construct as the digital revolution that is continually reinvigorated by ongoing events? And why has it persisted for so long and seems to keep going?

The digital revolution has acquired such wide acceptance and recognition because it is driven by various converging interests. On a political level, in almost all countries of the world, national governments and international organizations have drawn up projects and plans, betting on digitization and ensuring a large flow of public (and private) investments in this idea. In the name of the digital revolution, government agencies, ministries, commissions, and ad hoc working groups have been started, funds have been allocated to the military or those departments related to transport and communications, liberal policies have been adopted that have favored private initiative, as well as proprietary concentration and a level of flexibility in the labor market that has led to a state of persistent insecurity with regard to employment or income. A recent example is given by the post-pandemic economic recovery plans adopted by most countries across the globe (and the relative allocation of funds) in which "digital transformation" is not only mentioned obsessively but is also the necessary precondition for achieving all of the other objectives and, at the same time, is in itself a goal that must be pursued.

Digital companies, especially US but also Japanese and today increasingly Chinese and Indian, have been described and have described themselves as the protagonists of this revolution, churning out key concepts and mantras to support their socio-economic relevance. Clearly digital corporations are among the main beneficiaries of these one-dimensional narratives. The CEOs and entrepreneurs who founded them are gurus, capable even of guiding contemporary societies. Offices and factories are meccas to visit and contemplate in order to understand the future. Devices (whether they are watches, phones, or computers) become relics which you can use to take the revolution home or keep it in your pocket. So, the ideology of the revolution serves to make money, selling tangible products or intangible services, and to ensure that this pattern is repeated again in the future in an increasingly uncertain and volatile market: who wouldn't want to purchase (and regularly repurchase) a product that allows you to "join" the digital revolution, to enjoy it, to be part of it? There are a few lines from a 1994 article in *The New York Times* that offer some striking insight into this aspect:

> Once upon a time, the word "revolution" conjured up images of bloody peasant uprisings in China or of ragged guerrillas in Cuba overthrowing a dictatorship of the rich. A revolution was an angry upheaval of the power structure, a word usually associated with violence. But that's history. In America today, "revolution" has become a marketing buzzword most often invoked by buttoned-down business executives at telephone companies and movie studios and computer companies. They blather on and on about the information revolution and the communications revolution, and say, in effect, if you thought sliced bread was something, wait'll we really transform people's lives and in the process make lots and lots of money.

In a nutshell, the digital revolution has adopted and trivialized a term that has taken on a number of meanings throughout its history: *revolution*. In the "digital" world, revolution has become a brand and a catchphrase that, over time, has attracted similar concepts, such as innovation and disruption, which refer to the idea of radical change and linear progress. Revolutionary language, moreover, needs superlatives to remain electrifying. This trivialization can be seen in negative terms—even Lenin, writing for *Pravda* a few months before the October Revolution in 1917, expressed frustration at the unqualified use of the word "revolution" as "a gross deception of the people and of oneself"—but it is obviously strategic for large companies working in the sector. Not only and not so much because it allows private companies to amass rivers of money but above all because it positions them and their digerati managers as guides: they're the ones we can rely on and who will be able to show contemporary societies the way of the (strictly digital) future. Moreover, through the metaphor of revolution, digital corporations aim to extend their control over society and prevent any political body from attempting to regulate or limit the freedoms they have historically enjoyed. Any intervention that inhibits the expansion of digital companies is described as contrary to the digital revolution itself and, therefore, as an attempt to hinder an unstoppable global process.

"This 'digital revolution'—can we muscle in on that?"

Figure 5.1   How to profit of the digital revolution?

© cartoonstock.com. Cartoonist Bob Mankoff. Published in *The New Yorker*, January 25, 1999.

However, the digital revolution has also been useful for those who are not on the payroll of digital multinationals. Academics who have been predicting the advent of revolution or transformation for decades sell books and are interviewed by the media around the world (recently, an increasing number of those with an apocalyptic vision who theorize that the revolution will take a nefarious turn). Journalists write very popular columns about the revolution, often repeating the same concepts decades later or even using the same words for different objects. Then advertisers indulge in creating slogans that contain the term revolution, YouTubers make very popular tutorials on how to use digital relics, and even workers in sectors not connected to the digital world relate how their tasks have changed and, in the end, their lives as well thanks to or because of the digital revolution.

But thinking about it carefully, the great tale of the digital revolution is useful to virtually all human beings. Global consumers, all of us, have embraced the rhetoric of the revolution, have lined up in front of Apple stores to buy the new products that the revolution has constantly churned out. With blind confidence they have ordered all kinds of products from Amazon in the months of pandemic confinement or, more simply, they can no longer live without a smartphone, a computer, or a digital music player. They don't live without them because, in fact, these products fill life with increasingly personalized content and entertainment. Because they want to be part of the future, of a real or supposed global change, they embrace—or rather *we* embrace—the revolution uncritically (that is, suspending any degree of doubt) so as not to remain backward and be mocked by the younger generations for our inability to "turn on a computer." The narrative and poetics of the digital revolution is literally global: most people living on Earth buy or dream of buying the same digital devices, of being in step with the (revolutionary) times, of accessing the Internet in order to redeem themselves. The digital revolution justifies all the money spent on digital goods and services and the related expectations: it's not about throwing money away but rather about making indispensable investments in order to immerse oneself in the spirit of the times and prepare with confidence for the future.

The digital revolution, on the whole, is a powerful rhetorical and ideological construction in the name of which various social actors have invested time and attention as well as economic and symbolic capital. But it is not a "natural" construction. It has been patiently forged by the convergence of interests of all these actors to whom the idea of revolution serves to justify their actions and, ultimately, justify themselves: to justify investments, choices and industrial policies, their very existence on planet earth as human beings who connect (digitally) with others. As a spokesman for George W. Bush once said: "What made the digital revolution a revolution was an ambition to create its own reality," and this reality, patiently and socially constructed over time, is now *the* official reality.

But how did it come to thrive and why is it so enduring? First of all, as just mentioned, it is supported and embraced by billions of people and so it has simply become an idea that is as unconscious and automatic as it is popular and taken for granted. In part, this ideology emerged when others, of a political or religious nature, were waning and therefore represented an ideological lifeline. Moreover, it integrated perfectly with some of contemporary society's other one-dimensional ideas like capitalism, democracy, and the centrality of communication. Nothing is more reassuring than feeling like part of an epochal change that is, as we are continually reminded, decisive for humanity, unstoppable, and irresistible. So why resist slogans and formulations that are so seductive and easy to follow and which also add the thrill of the revolutionary dimension? After all, supporting the digital revolution means giving your life to a cause that doesn't require any particular commitment or sacrifices, that is not as rigid and prescriptive as a traditional religion, which is even fun, seductive, and salvific, as demonstrated during the COVID-19 pandemic.

The extraordinary endurance of the digital revolution ideology is also due to its metamorphic capacity. The revolution had few fixed points: some gurus, like Steve Jobs and Bill Gates, and maybe today Elon

Musk; some companies, such as GAFAM; some symbolic places, like Silicon Valley. Quite often it has changed names (or ways in which it is referred to), timelines, protagonists, objects (which inevitably age as soon as they're marketed), and the sanctuaries and regions of the world where it is sworn that the revolution is being made or will be made in the coming years have also changed. The digital revolution, despite a narrative that has lasted for decades, has changed its skin several times and this has always kept it fresh and, in fact, revolutionary.

But how much longer can this narrative last? Doesn't the digital revolution, which has been promising essentially the same things for several decades, risk being exposed, generating disillusionment and dissatisfaction among those who embraced it decades ago and still continue to believe in it today? Glimpsing the end of this great multi-voiced story isn't easy. Too many people, potentially all human beings, even those who have never used a digital device but dream of having one, are or believe themselves to be involved in its transformative logic and feel that they live in a digital age. But above all, it doesn't seem as if there is another ideology on the horizon that can adapt so easily to various cultures and fields of activity and is as engaging as that of the digital revolution. Perhaps the climate crisis may become so in the future, but at the moment there are still too many skeptics or heretics or political-business interests that oppose it. In short, the information revolution, the digital revolution and, increasingly, the digital transformation seem destined to dominate contemporary society's discourse for a long time to come.

Endnotes

On the digital revolution's capacity to influence political decisions, see Chandrasekhar, 2006. Regarding the economic and metaphorical relevance of the revolutionary discourse on private companies, see Pfaffenberger, 1988, and Tirosh and Schejter, 2017.

On corporate guru's profits, read Blondheim, 2009. According to Peppino Ortoleva (2019), on the one hand, the slogan "digital revolution" and the rhetoric of "nothing is like it was before" could wear thin from overuse, but on the other hand their popularity is constantly increasing (pp. 81–82). Other key contributions to the conclusion are Jenkins and Thorburn, 2003 (especially the introduction "The Digital Revolution, the Informed Citizen, and the Culture of Democracy"); Kleine and Unwin, 2009; Kortti, 2021; Schiller, 1983. There are three quotations in the conclusion. The one from *The New York Times* comes from Andrews, 1994, p. 1; the second is found in Lenin's essay "A Mote in the Eye" (1964), the third from the spokesman of George W. Bush can be found in Johns, 2012, p. 861.

References

Agar, Nicholas. (2019). *How to be Human in the Digital Economy*. The MIT Press.

Albrecht, Karl. (2014, March 1). The Information Revolution's Broken Promises. *Futurist, 48(2)*.

Alexander, Jeffrey. (1990). The Sacred and Profane Information Machine: Discourse about the Computer as Ideology. *Archives de Sciences Sociales des Religions, 35(69)*, 161–171.

Alter, Adam. (2017). *Irresistible: The Rise of Addictive Technology and the Business of Keeping Us Hooked*. Penguin.

Andrews, Edmund L. (1994, October 30). Changing the Wiring Takes Time: Installing All the New Wiring Takes Time. *The New York Times*, p. 1.

Anguiano, Dani. (2022, November 15). Steve Jobs' Old Birkenstocks Sell for Nearly $220,000. *The Guardian* online.

Anthes, Gary. (2006, April 3). Bits to Atoms (and Atoms to Bits). *Computerworld*, 40.

Are We Aware We Are Living through a Digital Revolution? (2012, February 12). *Professional Adviser*.

Arendt, Hannah. (1990). *On Revolution*. Penguin.

Attali, Jacques. (1992). *Millennium: Winners and Losers in the Coming World Order*. Times Books.

Babinet, Gilles. (2014). *L'ère numérique, un nouvel âge de l'humanité*. Le Passeur.

Babinet, Gilles. (2016). *Transformation digitale*. Le Passeur.

Balbi, Gabriele, and Magaudda, Paolo. (2018). *A History of Digital Media. An Intermedia and Global Perspective*. Routledge.

Balbi, Gabriele, Ribeiro, Nelson, Schafer, Valérie, and Schwarzenegger, Christian. (2021). "Digging into Digital Roots. Towards a Conceptual Media and Communication History". In Gabriele Balbi, Nelson Ribeiro, Valérie Schafer, and Christian Schwarzenegger (Eds.), *Digital Roots. Historicizing Media and Communication Concepts of the Digital Age* (pp. 1–16). De Gruyter.

Ball, Matthew. (2022). *The Metaverse: and How It Will Revolutionize Everything*. Liveright Publishing Corporation.

Barber, John T. (2006). *The Black Digital Elite: African American Leaders of the Information Revolution*. Praege.

Barbrook, Richard. (2007). *Imaginary Futures: From Thinking Machines to the Global Village*. Pluto.

Barbrook, Richard, and Cameron, Andy. (1996). The Californian Ideology. *Science as Culture*, 6(1), 44–72.

Barlow, John P. (1995). Is There a There in Cyberspace? *Utne Reader*, March-April, 50–56.

Barnatt, Christopher. (2001). The Second Digital Revolution. *Journal of General Management*, 27(2), 1–16.

Barris, Michael. (2014, May 1). China "to Lead" Digital Revolution. *China Daily*.

Beaumont, Peter. (2011, February 25). The Truth about Twitter, Facebook and the Uprisings in the Arab World. *The Guardian* online.

Behringer, Wolfgang. (2006). Communications Revolutions: A Historiographical Concept. *German History*, 24(3), 333–374.

Bello, Alessandro, Blowers, Tonya, Schneegans, Susan, and Straza, Tiffany. (2021). To Be Smart, the Digital Revolution Will Need to Be Inclusive. *UNESCO Science Report: The Race against Time for Smarter Development*. UNESCO. Available at https://unesdoc.unesco.org/ark:/48223/pf0000375429.

Beniger, James R. (1986). *The Control Revolution: Technological and Economic Origins of the Information Society*. Harvard University Press.

Blondheim, Menahem. (2009). Narrating the History of Media Technologies. In Michael Bailey (Ed.), *Narrating Media History* (pp. 212–227). Routledge.

Borpuzari, Pranbihanga, and Bhattacharya, Ashutosh. (2016, June 16). India's Digital Revolution Similar to Europe's Industrial Revolution: Droom's Sandeep Aggarwal. *The Economic Times*.

Bory, Paolo. (2020). *The Internet Myth: From the Internet Imaginary to Network Ideologies*. University of Westminster Press.

Bourdon, Jérôme. (2018). The Case for the Technological Comparison in Communication History. *Communication Theory*, 28(1), 89–109.

Bourdon, Jérôme, and Balbi, Gabriele. (2021). Questioning (Deep) Mediatization: A Historical and Anthropological Critique. *International Journal of Communication*, 15, 807–2826.

Bowers, Chet. (2014). Is the Digital Revolution Driven by an Ideology? *Studies in Sociology of Science*, 5(3), 169–178.

Boym, Svetlana. (2001). *The Future of Nostalgia*. Basic Books.

Briscoe, Bob, Odlyzko, Andrew, and Tilly, Benjamin. (2016, July 1). Metcalfe's Law is Wrong. *IEEE Spectrum*.

Brock, Gerald W. (2003). *The Second Information Revolution*. Harvard University Press.

Bronson, Po. (1998, January). Is the Revolution Over? Report from Ground Zero: Silicon Valley. *Wired*.

Brzezinski, Zbigniew. (1970). *Between Two Ages. America's Role in the Technetronic Era*. Viking Press.

Business Monthly: Commentators Talk of a Bright Future for the Digital Revolution. (2002, November 2). *The Independent*.

Butler, Octavia. (2009). *Kindred*. Beacon.

Cairncross, Frances. (1997). *The Death of Distance: How the Communications Revolution Will Change Our Lives*. Harvard Business School Press.

Campbell, Heidi A., and La Pastina, Antonio C. (2010). How the iPhone Became Divine: New Media, Religion and the Intertextual Circulation of Meaning. *New Media and Society*, 12(7), 1191–1207.

Carey, James W. (1989). *Communication as Culture: Essays on Media and Society*. Unwin Hyman.

Carey, John, and Elton, Martin C. J. (2018). *When Media Are New: Understanding the Dynamics of New Media Adoption and Use*. University of Michigan Press.

Casaleggio, Gianroberto. (2004). *Web Ergo Sum*. Sperling & Kupfer.

Castells, Manuel. (2001). *Internet Galaxy*. Oxford University Press.

Chandrasekhar, C.P. (2006). Who Needs a "Knowledge Economy": Information, Knowledge and Flexible Labour. *Social Scientist*, 34(1/2), 70–87.

Chirinos, Carmela. (2022, March 24). Elon Musk's Wildest Predictions about the Future, Some of Which Have Come True. *Fortune*.

Cohan, Al S. (1975). *Theories of Revolution: An Introduction*. Wiley.

Collard, Neil. (2021, February 26). What Karl Marx and Jeff Bezos Can Teach Us About Digital Transformation. *ceotodaymagazine.com*.

Collato, Federica. (2010). Is Bangalore the Silicon Valley of Asia? *Journal of Indian Business Research*, 2(1), 52–65.

Cormack, Michael J. (1992). *Ideology*. University of Michigan Press.

Cortada, James W. (2012). *The Digital Flood: The Diffusion of Information Technology Across the U.S., Europe, and Asia*. Oxford University Press.

Crawford, Kate. (2021). *Atlas of AI: Power, Politics, and the Planetary Costs of Artificial Intelligence*. Yale University Press.

Crisp, Jason (1983, October 25). Optical Fibres Make a Commercial Breakthrough. *Financial Times*.

Curran, James. (2012). Rethinking Internet History. In James Curran, Natalie Fenton, and Des Freedman (Eds.), *Misunderstanding the Internet* (pp. 34–65). Routledge.

Dass, Vishwas. (2020, January 26). India to Take the Lead in Digital Revolution, Says Gujarat Energy Minister Saurabhbhai Patel. *Express Computer*.

Daub, Adrian. (2020). *What Tech Calls Thinking: An Inquiry into the Intellectual Bedrock of Silicon Valley*. Farrar, Straus and Giroux.

Davies, Russell. (2009, June 12). What More Is There to Say? The Digital Revolution Is Over. *Campaign*.

De Biase, Luca. (2003). *Edeologia. Critica del fondamentalismo digitale*. Laterza.

de Sola Pool, Ithiel. (1982). The Culture of Electronic Print. *Daedalus, 111*(4), 17–31.

de Tocqueville, Alexis. (2004). *Democracy in America, the Complete and Unabridged Volumes I and II*. Bantam Dell.

Dewar, James A. (1998). *The Information Age and the Printing Press: Looking Backward to See Ahead*. RAND Corporation.

Digital Transformation: History, Present, and Future Trends. (2016). https://auriga.com/blog/2016/digital-transformation-history-present-and-future-trends/.

Dobbin, Ben. (1996, December 9). Fiber Optics Driving Information Revolution. *Austin American-Statesman*.

Dong-woo, Nam. (2006, September 3). Samsung Positions Firm for Digital Boom in 2010. *Korea JoongAng Daily*.

Doug Engelbart. Obituaries Computer Visionary Who Foresaw the Information Revolution and Invented the Desktop Mouse. (2013, July 8). *The Daily Telegraph*.

Driessens, Olivier. (2023). Not Everything Is Changing: On the Relative Neglect and Meanings of Continuity in Communication and Social Change Research. *Communication Theory, 33*(1), 32–41.

Drucker, Peter F. (1968). *The Age of Discontinuity. Guidelines to Our Changing Society*. Harper & Row.

Eisenstein, Elizabeth L. (1993). *The Printing Revolution in Early Modern Europe*. Cambridge University Press.

Eisenstein, Elizabeth L. (2002). An Unacknowledged Revolution Revisited. *The American Historical Review, 107*(1), 87–105.

Elliott, Heidi. (1997, December). Happy 50th. *Electronic Business*.

Ensmenger, Nathan. (2012). The Digital Construction of Technology: Rethinking the History of Computers in Society. *Technology and Culture, 53*(4), 753–776.

Epstein, Jason. (2008). The End of the Gutenberg Era. *Library Trends, 57*(1), 8–16.

Essadki, Jawad. (2015). *The Manifesto of the Digital Revolution: Reflections, Observations & Forecast*. Self-published.

Etzo, Sebastiana, and Collender, Guy. (2010). The Mobile Phone "Revolution" in Africa: Rhetoric or Reality? *African Affairs, 109*(437), 659–668.

European Green Digital Coalition. (2020). Declaration to support the Green and Digital Transformation of the European Union. https://www.greendigitalcoalition.eu/declaration/#:~:text=The%20Declaration%20to%20support%20the,the%20fight%20against%20climate%20change.

Farrington, Gregory C. (1996). ENIAC: The Birth of the Information Age. *Popular Science, 248*(3): 74.

Fickers, Andreas. (2012). The Emergence of Television as a Conservative Media Revolution: Historicising a Process of Remediation in the PostWar Western European Mass Media Ensemble. *Journal of Modern European History, 10*(1), 49–75.

Fischer, Hervé. (2006). *Digital Shock: Confronting the New Reality.* McGill-Queen's University Press.

Floridi, Luciano. (2014). *The Fourth Revolution. How the Infosphere is Reshaping Human Reality.* Oxford University Press.

Forester, Tom (Ed.). (1981). *The Microelectronics Revolution.* The MIT Press.

Foster, John Bellamy. (2020). *The Return of Nature: Socialism and Ecology.* Monthly Review Press.

Freeman, Chris, and Louçã, Francisco. (2001). *As Time Goes By. From the Industrial Revolution to the Information Revolution.* Oxford University Press.

Friedrich, Otto. (1983, January 4). The Computer. *Time.*

Fuchs, Christian. (2018). Industry 4.0: The Digital German Ideology. *TripleC, 16*(1), 280–289.

Fuchs, Christian. (2020). *Nationalism on the Internet: Critical Theory and Ideology in the Age of Social Media and Fake News.* Routledge.

Fujitsu. A Survey of Telecommunications. (1987, October 17). *The Economist,* 21–22.

G20. (2021, January 12). *Economic Recovery Will Be Through Digital Revolution.* http://www.g20italy.org/g20-2021-economic-recovery-will-be-through-digital-revolution.html.

Gabrys, Jennifer. (2011). *Digital Rubbish. A Natural History of Electronics.* University of Michigan Press.

Garner, Rochelle. (2016, March 22). Andy Grove, "Intel Cofounder and Chip Visionary, Has Died." *CNET News.com,* 22 March 2016.

Gates, Bill. (1995). *The Road Ahead.* Viking.

Gates, Bill. (2008, January 24). Perpetual (Digital) Revolution. *The Wall Street Journal.*

Gates, Bill. (2023, March 21). The Age of AI has Begun. Artificial Intelligence is as Revolutionary as Mobile Phones and the Internet. *Gates Notes.* https://www.

gatesnotes.com/The-Age-of-AI-Has-Begun?WT.mc_id=20230321100000_
Artificial-Intelligence_BG-TW_&WT.tsrc=BGTW.

Gates Sceptical of Google's Balloon Internet Connectivity Helping the Poor. (2013, August 12). *Domain-B.com.*

Getov, Vladimir. (2013). Computing Laws: Origins, Standing, and Impact. *Computer, 46*(12), 24–25.

Ghosh, Jayanta. (2003, September 3). It's the Start of Digital Era. *The Times of India.*

Gigaphoton Announces New Corporate Slogan: "The Future Is Today". (2019, August 1). *Businesswire.com.*

Gillies, James, and Cailliau, Robert. (2000). *How the Web Was Born: The Story of the World Wide Web.* Oxford University Press.

Gohring, Nancy. (Ed.) (2022). *IDC FutureScape: Worldwide Future of Digital Innovation 2023 Predictions.* https://www.idc.com/eu/events/70567-idc-future-of-digital-innovation?g_clang=ENG.

Gordon, Robert J. (2016). *The Rise and Fall of American Growth. The U.S. Standard of Living Since the Civil War.* Princeton University Press.

Graham-Cumming, John. (2012, May 30). Alan Turing: Computation. *New Scientist.*

Gramsci, Antonio. (2011). *Prison Notebooks,* translated by J.A. Buttigieg and A. Callari. Columbia University Press.

Gullard, Pamela. (1981). The Charles Babbage Institute for the History of Information Processing (CBI). *Isis, 72*(2), 262–264.

Gunnerson, Ronnie. (1993). The Digital Revolution Is a Fait Accompli. *Dealerscope, 35*(6), 70.

Hart, Keith. (2009). An Anthropologist in the World Revolution. *Anthropology Today, 25*(6), 24–25.

Hartley, Steven. (2016, April 11). Get Ready for the Second Digital Revolution. *Telecom Asia Online.*

Headrick, Daniel R. (2000). *When Information Came of Age: Technologies of Knowledge in the Age of Reason and Revolution, 1700–1850.* Oxford University Press.

Hepp, Andreas. (2020). *Deep Mediatization.* Routledge.

Hinkson, John. (2013–2014). Why Do We Place Our Hope in Technology? A Secular Faith? *Arena Journal, 41/42,* 59–92.

Hobsbawm, Eric J. (1986). Revolution. In Roy Porter and Mikuláš Teich (Eds.), *Revolution in History* (pp. 5–46). Cambridge University Press.

Humphreys, Peter. (1986). Legitimating the Communications Revolution: Governments, Parties and Trade Unions in Britain, France and West Germany. *West European Politics, 9*(4), 163–194.

Hundley, Richard O., Anderson, Robert H., Bikson, Tora K., and Neu, C. Richard. (2003). *The Global Course of the Information Revolution: Recurring Themes and Regional Variations*. RAND Corporation.

India at the Cusp of Digital Revolution, Says PM Narendra Modi. (2018, February 20). *The Financial Express*.

Information Revolution Is Only Just Starting, Warns Expert. (2003, May 7). *Birmingham Post*.

Intel Celebrates 40 Years of Digital Revolution. (2011, November 15). *Business Wire*.

Isaacson, Walter. (1997, December 29). Man of the Year: Andrew Grove. *Time*.

Isaacson, Walter. (2011a). *Steve Jobs*. Simon & Schuster

Isaacson, Walter. (2011b). *The Innovators: How a Group of Inventors, Hackers, Geniuses, and Geeks Created the Digital Revolution*. Simon & Schuster.

Itoh, Tsunatoshi. (2001). Silicon Island Kyushu's Integrated Circuit Industry during the 1990s. *Annals of the Society for Industrial Studies, 16*, 57–71.

Jackson, Tim. (1997). *Inside Intel: Andy Grove and the Rise of the World's Most Powerful Chip Company*. Dutton.

James, Matt. (2019, blog). *How to Keep Faith in the Digital Revolution*. https://care.org.uk/news/2019/05/how-to-keep-the-faith-in-the-digital-revolution.

Jedlowski, Paolo. (2017). *Memorie del futuro. Un percorso tra sociologia e studi culturali*. Carocci.

Jemisin, Nora Keita. (2020). *The City We Became*. Hachette.

Jemisin, Nora Keita. (2022). *The World We Made*. Orbit.

Jenkins, Henry, and Thorburn, David. (2003). *Democracy and New Media*. The MIT Press.

Jenkins, Simon. (2012, July 10). For the Digital Revolution, This Is the Robespierre Moment. *The Guardian*, p. 30.

John, Richard R. (1994). American Historians and the Concept of the Communications Revolution. In Lisa Bud-Frierman (Ed.), *Information Acumen: The Understanding and Use of Knowledge in Modern Business* (pp. 98–110). Routledge.

Johns, Adrian. (1998). *The Nature of the Book: Print and Knowledge in the Making*. University of Chicago Press.

Johns, Adrian. (2002). How to Acknowledge a Revolution. *The American Historical Review, 107*(1), 106–125.

Johns, Adrian. (2012). Gutenberg and the Samurai: Or, the Information Revolution Is History. *Anthropological Quarterly, 85*(3), 859–883.

Johnson, Malcolm. (2019, June 18). Let's Align to Make the Digital Revolution a Development Revolution. ITU website. https://www.itu.int/hub/2020/02/lets-align-to-make-the-digital-revolution-a-development-revolution/.

Karns Alexander, Jennifer. (2020). Introduction: The Entanglement of Technology and Religion. *History and Technology*, *36*(2), 165–186.

Karpf, David. (2018, October 1). The Future Was So Delicious, I Ate It All. *Wired*.

Kaufman, Micha (2012, October 5). The Internet Revolution Is the New Industrial Revolution. *Forbes*.

Keegan, Paul. (1995, May 21). The Digerati! Wired Magazine Has Triumphed by Turning Mild-Mannered Computer Nerds into a Super-Desirable Consumer Niche. *The New York Times*.

Keegan, Victor. (1993, March 22). Economics Notebook: Wake Up! Beyond Dogma Lies the Digital Revolution. *The Guardian*, p. 10.

Keegan, Victor. (1996, August 1). The Everything Machine. *The Guardian*.

Keju, Wang. (2018, December 24). China Leads Global Net Revolution. *China Daily*.

Kleine Dorothea, and Unwin, Tim. (2009). Technological Revolution, Evolution and New Dependencies: What's New about ICT4D? *Third World Quarterly*, *30*(5), 1045–1067.

Knowles, Tom. (2021, July 13). AI Will Have a Bigger Impact than Fire, Says Google Boss Sundar Pichai. *The Times*.

Koch Honored as a "True Hero" of the Information Revolution. (1999, June 4). *Business Wire*.

Kortti, Jukka. (2021). Revolution Talk and Media History. *Academia Letters*, *811*, 1–5.

Koselleck, Reinhart. (2004). *Futures Past. On the Semantics of Historical Time*. Columbia University Press.

Köstlbauer, Josef. (2011, April 17). *Digitale Revolution*. Paper presented at the Historikertagung des Instituts für Österreichkunde, St. Poelten.

Kroker, Arthur, and Kroker, Marilouise. (1999). Digital Ideology: E-Theory (1). *CTHEORY*, September.

Krugman, Paul. (2023, April 4). The Internet Was an Economic Disappointment. *The New York Times*.

Krzywdzinski, Martin, Gerber, Christine, and Evers, Maren. (2018). The Social Consequences of the Digital Revolution. In Pietro Grasso and Giuliana Chiaretti (Eds.), *Le grandi questioni sociali del nostro tempo. A partire da Luciano Gallino* (pp. 101–120). Ca' Foscari University Press.

Kuhn, Thomas S. (1970). *The Structure of Scientific Revolutions*. The University of Chicago Press.

Kulwiec, Ray. (1994). You Are Living in a Revolution (Information Revolution). *Modern Materials Handling*, *49*(101).

Kushairi, Ahmad. (1995, November 6). Information Revolution Important for Progress. *The New Straits Times*.

Lane, M. Raymond. (1997, March 1). The Second Information Revolution. *New Perspectives Quarterly*, 14(2).

Lee, Jean H. (2011, July 25). Quiet Digital Revolution Under Way in North Korea. *The Sydney Morning Herald*.

Lenin, Vladimir Ilyich. (1964). A Mote in the Eye. In *Lenin Collected Works* (pp. 565–567). Progress Publishers, Volume 24.

Lenk, Kurt. (1973). *Theorien der Revolution*. W. Fink.

Levinson, Paul. (1993, July-August). Five Views of St. Marshall. *Wired*.

Levy, Steven. (2010). *Hackers: Heroes of the Computer Revolution*. O'Reilly Media.

Li, Xing. (2006, May 16). Looking to Our High-Tech Future. *China Daily North American*, 13.

Little, Ben, and Winch, Alison. (2021). *The New Patriarchs of Digital Capitalism. Celebrity Tech Founders and Networks of Power*. Routledge.

Lüthy, Christoph. (2000). Caught in the Electronic Revolution. Observations and Analyses by Some Historians of Science, Medicine, Technology, and Philosophy. *Early Science and Medicine*, 5(1), 64–92.

Machlup, Fritz. (1962). *The Production and Distribution of Knowledge in the United States*. Princeton University Press.

MacKinnon, Angus. (1995, February 25). Markets Are the Bolsheviks of the Information Revolution. *Agence France-Presse*.

Madrick, Jeff. (2014). The Digital Revolution That Wasn't. *Harper's Magazine*, 328(1964), 11–13.

Malone, Michael S. (1998, August). Forget Digital Utopia . . . We Could Be Headed for Technofascism. *Upside*.

Margolis, Michael, and Resnick, David. (2000). *Politics as Usual: The Cyberspace "Revolution"*. Sage.

Markoff, John. (1992, January 29). Business Technology; Pools of Memory, Waves of Dispute. *The New York Times*.

Markoff, John. (2008, March 13). Joseph Weizenbaum, Famed Programmer, Is Dead at 85. *The New York Times*.

Marsh, Allison C. (2013). Revolution: The First 2,000 Years of Computing: The Computer History Museum, Mountain View, California. *Technology and Culture*, 54(3), 640–649.

Marvin, Carolyn. (1990). *When Old Technologies Were New: Thinking About Electric Communication in the Late Nineteenth Century*. Oxford University Press.

Marx, Karl. (1976). *The Class Struggles in France (1848–1850)*. International Publishers.

Marx, Leo. (1964). *The Machine in the Garden*. Oxford University Press.

Marx, Leo. (2010). Technology: The Emergence of a Hazardous Concept. *Technology and Culture, 51*(3), 561–577.

Masuda, Yoneji. (1980). *The Information Society as Post-Industrial Society*. Institute for the Information Society.

Mathews, Jessica T. (2000). The Information Revolution. *Foreign Policy, 119,* 63–65.

McChesney, Robert. (2007). *Communication Revolution: Critical Junctures and the Future of Media*. The New Press.

McGann, Jerome. (2005). Culture and Technology: The Way We Live Now, What Is to Be Done? *New Literary History, 36*(1), 71–82.

McLellan, David. (1995). *Ideology*. University of Minnesota Press.

McLuhan, Marshall. (1964). *Understanding Media: The Extensions of Man*. McGraw-Hill.

Merleau-Ponty, Maurice. (1968). *The Visible and the Invisible*. Northwestern University Press.

Merriam Webster. https://www.merriam-webster.com/dictionary/.

Merrit, Bob. (2016). *The Digital Revolution*. Morgan & Claypool.

More, Thomas. 2003. *Utopia*. Penguin.

Morin, Edgar. (1962). *L'esprit du temps: essai sur la culture de masse*. Grasset.

Morozov, Evgeny. (2014). *To Save Everything, Click Here: Technology, Solutionism, and the Urge to Fix Problems That Don't Exist*. Penguin.

Mosco, Vincent. (2004). *The Digital Sublime: Myth, Power, and Cyberspace*. The MIT Press.

Mosco, Vincent, and Herman, Andrew. (1980). Communication, Domination and Resistance. *Media, Culture and Society, 2*(4), 351–365.

Mounier, Pierre. (2018). *Les humanités numériques: une histoire critique*. Éditions de la Maison des sciences de l'homme.

Muñoz, José Esteban. (2019). *Cruising Utopia, 10th Anniversary Edition: The Then and There of Queer Futurity*. New York University Press.

Musso, Pierre. (2003). *Critique des réseaux*. Presses universitaires de France.

Musso, Pierre. (2017). Network Ideology: From Saint-Simonianism to the Internet. In José Luís Garcia (Ed.), *Pierre Musso and the Network Society* (pp. 19–66). Springer International Publishing.

Naisbitt, John. (1984). Megatrends: The New Directions Transforming Our Lives. *Futura*, p. 28.

Natale, Simone, and Pasulka, Diana W. (Eds.) (2020). *Believing in Bits: Digital Media and the Supernatural*. Oxford University Press.

Naughton, John. (2014). *From Gutenberg to Zuckerberg: Disruptive Innovation in the Age of the Internet*. Quercus.

Negro Gianluigi, and Wu, Jing. (2020). Exporting the Silicon Valley to China. *Online Journal of Communication and Media Technologies*, *10*(3), 1–17.

Negroponte, Nicholas. (1995). *Being Digital*. Knopf.

New On-Line Service Unveiled by Singapore. (1995, 28 March). *Asian Wall Street Journal*, p.14.

Noble, David F. (1997). *The Religion of Technology: The Divinity of Man and the Spirit of Invention*. A.A. Knopf.

Nora, Simon, and Minc, Alain. (1978). *L'informatisation de la société*. La documentation française.

Nye, David E. (1990). *Electrifying America: Social Meanings of a New Technology, 1880–1940*. MIT Press.

Nye, David E. (1994). *The American Technological Sublime*. MIT Press.

O'Lemmon, Matthew. (2022). The Worst Mistake 2.0? The Digital Revolution and the Consequences of Innovation. *AI & Society*.

Ortoleva, Peppino. (2019). *Miti a bassa intensità: racconti, media, vita quotidiana*. Einaudi.

Oxford Learner's Dictionaries. https://premium.oxforddictionaries.com/definition/american_english/.

Paul Allen. Obituary. (2018, October 17). *The Daily Telegraph*.

Perez, Carlota. (2004). Technological Revolutions, Paradigm Shifts and Socio-Institutional Change. In Erik S. Reinert (Ed.), *Globalization, Economic Development and Inequality: An Alternative Perspective* (pp. 217–242). Edward Elgar.

Pfaffenberger, Bryan. (1988). The Social Meaning of the Personal Computer: Or, Why the Personal Computer Revolution Was No Revolution. *Anthropological Quarterly*, *61*(1), 39–47.

Pitron, Guillaume. (2021). *L'enfer numérique: voyage au bout d'un like*. Éditions Les Liens.

Pollock, Neil, and Williams, Robin. (2016). *How Industry Analysts Shape the Digital Future*. Oxford University Press.

Porat, Marc Uri. (1977). *The Information Economy: Definition and Measurement*. United States Department of Commerce.

Porter, Roy. (1986). The Scientific Revolution: A Spoke in the Wheel? In Roy Porter and Mikuláš Teich (Eds.), *Revolution in History* (pp. 290–316). Cambridge University Press.

Pugh O'Mara, Margaret. (2019). *The Code: Silicon Valley and the Remaking of America*, Penguin Press.

Ray Kurzweil Reader. (n.d.). *A Collection of Essays by Ray Kurzweil published on KurzweilAI.net 2001–2003*. http://www.kurzweilai.net/pdf/RayKurzweilReader.pdf.

Rehmann, Jan. (2007). Ideology Theory. *Historical Materialism*, 15(4), 211–239.

Report of the Special Commission of the Swiss National Council. (2000, May, 29). Legislature Program 1999–2003. https://www.fedlex.admin.ch/eli/fga/2000/1289/it.

Robinson, Brian. (2022, February 1). Remote Work Is Here To Stay And Will Increase into 2023. *Forbes*.

Robinson, Kim Stanley. (2020). The Ministry for the Future. Orbit.

Rogers, Adam. (2018, October 1). We Predicted a Digital Revolution with All the Fervor of True Believers. Then the Revolution Conquered All. *Wired*.

Rossetto, Louis. (1993). The Wired Manifesto. *Wired*, March-April.

Rossi-Landi, Ferruccio. (2005). *Ideologia. Per l'interpretazione di un operare sociale e la ricostruzione di un concetto*. Meltemi.

Rud, Anthony G. Jr. (1997). Review: Being Digital by Nicholas Negroponte. *Educational Researcher*, 26(7).

Russell, Andrew L. (2017). Hagiography, Revisionism & Blasphemy in Internet Histories. *Internet Histories*, 1(1-2), 15–25.

Russia a Leader in the Information Revolution. (2013, April 11). *The Moscow Times*.

Sadin, Éric. (2016). *La silicolonisation du monde: l'irrésistible expansion du libéralisme numérique*. Éditions l'Échappée.

Sample, Ian. (2015, February 13). Google Boss Warns of 'Forgotten Century' with Email and Photos at Risk. *The Guardian*.

Sarikakis, Katharine, and Thussu, Daya. (2006). *Ideologies of the Internet*. Hampton Press.

Savic, Bob. (2020, June 19). China's New Digital Industrial Transformation. *The Diplomat*.

Scardigli, Victor. (1988). Towards Digital Man? *Design Issues*, 4(1/2), 152–167.

Schallmo, Daniel, Williams, Christopher A., and Boardman, Luke. (2017). Digital Transformation of Business Models. Best Practices, Enablers, and Roadmap. *International Journal of Innovation Management*, 21(8).

Schiller, Herbert I. (1983). The Communications Revolution: Who Benefits? *Media Development*, 4, 18–20.

Schwarzenegger, Christian, and Balbi, Gabriele. (2020). When the "Messiah" Went to "Mecca": Envisioning and Reporting the Digital Future at the CeBIT Tech Fair (1986–2018). *Convergence*, 26(4), 716–731.

Scolari, Carlos A. (2023). *On the Evolution of Media. Understanding Media Change.* Routledge.

Shachtman, Noah. (2002, May). Honoring Wired's Patron Saint. *Wired.*

Sisto, Davide. (2020). *Online Afterlives: Immortality, Memory, and Grief in Digital Culture.* The MIT Press.

Supp-Montgomerie, Jenna. (2021). *When the Medium Was the Mission: the Atlantic Telegraph and the Religious Origins of Network Culture.* New York University Press.

Swerdlow, Joel L. (1995). Information Revolution. *National Geographic, 188*(4), 16–17.

Syvertsen, Trine. (2020). *Digital Detox: The Politics of Disconnecting.* Emerald.

The Compact Edition of the Oxford English Dictionary (1976). Volume I and Volume II, Oxford University Press.

The Epicentre of the Digital Revolution. (2014, October 20). *The Financial Express.*

The Future Has Arrived—It's Just Not Evenly Distributed Yet. (2021). *Quote Investigator.* https://quoteinvestigator.com/2012/01/24/future-has-arrived.

The Information Revolution is Coming. (2003, May 13). *A&G Information Services: Comtex.*

Tirosh, Noam, and Schejter, Amit. (2017, June 13). Who Benefits from the So-Called New Media Revolution? In *Digital Society Blog. Dossier How Metaphors Shape the Digital Society.* https://www.hiig.de/en/who-benefits-from-the-so-called-new-media-revolution/.

Toffler, Alvin. (1980). *The Third Wave.* William Morrow.

Toffler, Alvin, and Toffler, Heidi. (2001, March 29). New Economy? You Ain't Seen Nothin' Yet. *The Wall Street Journal.* March 29, p. 14.

Tremblay, Gaetan. (1995). The Information Society: from Fordism to Gatesism. *Canadian Journal of Communication, 20*(4), 461–482.

Tsuria, Ruth. (2021). Digital Media: When God Becomes Everybody. The Blurring of Sacred and Profane. *Religions, 12*(110), 1–12.

Tuomi, Ilkka. (2002). The Lives and Death of Moore's Law. *First Monday, 7*(11).

United States. National Telecommunications Information Administration. (1993). *The National Information Infrastructure: Agenda for Action.* US Dept. of Commerce, National Technical Information Service. https://clintonwhitehouse6.archives.gov/1993/09/1993-09-15-the-national-information-infrastructure-agenda-for-action.html.

Vitalis, André. (2016). *The Uncertain Digital Revolution.* ISTE-Wiley.

Waters, Richard. (2014, October 31). FT Interview With Google Co-Founder and CEO Larry Page. *Financial Times.*

Weber, Max. (2002) The Protestant Ethic and the "Spirit" of Capitalism and Other Writings (P. Baehr and G. C. Wells, Trans.). Penguin.

Weinberger, David. (2007). *Everything Is Miscellaneous. The Power of the New Digital Disorder*. Times Books,

Weizenbaum, Joseph. (1976). *Computer Power and Human Reason: From Judgment to Calculation*, W.H. Freeman.

Wiener, Norbert. (1948). *Cybernetics; or, Control and Communication in the Animal and the Machine*. MIT Press.

Wiener, Norbert. (1989). *The Human Use of Human Beings: Cybernetics and Society*. Free Associations Books.

Williams, Frederick. (1982). *The Communications Revolution*. Sage.

Winner, Langdon. (1984). Mythinformation in the High-Tech Era. *Bulletin of Science, Technology & Society*, 4(6), 582–596.

Winston, Brian. (1998). *Media Technology and Society. A History: From the Telegraph to the Internet*. Routledge.

Winters, Paul A. (1998). *The Information Revolution: Opposing Viewpoints*. Greenhaven Press.

Wolf, Gary. (1996a, January). Channeling McLuhan: The Wired Interview with the Magazine's Patron Saint. *Wired*.

Wolf, Gary. (1996b, February). The Wisdom of Saint Marshall, the Holy Fool. *Wired*.

Wolf, Gary. (2003). *Wired. A Romance*. Random House.

The World Economic Forum (2017). *Digital Transformation Initiative in Collaboration with Accenture. Unlocking $100 Trillion for Business and Society from Digital Transformation*. https://digiwisehub.com/download/accenture-dti-executive-summary.pdf

YouTube. (2008). *Gaia. The Future of Politic*.

Yu'an, Zhang. (1994, October 7). Country Embracing Information Revolution. *China Daily*.

Zacher, Lech W. (2015). Digital Future(s). *Encyclopedia of Information Science and Technology*. IGI Global.

Zuckerberg, Mark. (2014, July 7). Mark Zuckerberg on a Future Where the Internet Is Available to All. *The Wall Street Journal* online. https://www.wsj.com/articles/mark-zuckerberg-on-a-future-where-the-internet-is-available-to-all-1404762276.

Zysman, John, and Newman, Abraham. (Eds.) (2006). *How Revolutionary Was the Digital Revolution? National Responses, Market Transitions, and Global Technology*. Stanford Business Books.

Index

For the benefit of digital users, indexed terms that span two pages (e.g., 52–53) may, on occasion, appear on only one of those pages.